# Industrial
# Network Security
## Second Edition

AF148792

# Industrial
# Network Security
## Second Edition

By David J. Teumim

Printed in the United States of America.
10 9 8 7 6 5 4

ISBN 978-1-936007-07-3

ISA
67 Alexander Drive
P.O. Box 12277
Research Triangle Park, NC 27709
www.isa.org

**Library of Congress Cataloging-in-Publication Data**
Teumim, David J.
  Industrial network security / By David J. Teumim. -- 2nd ed.
      p. cm.
  ISBN 978-1-936007-07-3 (pbk.)
  1. Computer networks--Security measures. 2. Computer security. I. Title.
  TK5105.59.T48 2010
  005.8--dc22
                                                    2009049791

Notice
    The information presented in this publication is for the general education of the reader. Because neither the author nor the publisher have any control over the use of the information by the reader, both the author and the publisher disclaim any and all liability of any kind arising out of such use. The reader is expected to exercise sound professional judgment in using any of the information presented in a particular application.
    Additionally, neither the author nor the publisher have investigated or considered the effect of any patents on the ability of the reader to use any of the information in a particular application. The reader is responsible for reviewing any possible patents that may affect any particular use of the information presented.
    Any references to commercial products in the work are cited as examples only. Neither the author nor the publisher endorse any referenced commercial product. Any trademarks or tradenames referenced belong to the respective owner of the mark or name. Neither the author nor the publisher make any representation regarding the availability of any referenced commercial product at any time. The manufacturer's instructions on use of any commercial product must be followed at all times, even if in conflict with the information in this publication.

# Acknowledgments

My appreciation is expressed for the people who helped and inspired me to write the second edition of this book.

Once again, my special thanks go to my ISA editor, Susan Colwell.

John Clem, from Sandia National Laboratories, contributed content on Red Teaming for the new Chapter 9, New Topics in Industrial Network Security.

My good friend from college, Andy Hagel, provided content and review for Chapter 3, COTS and Connectivity.

As with the first edition, Dave Mills of Procter & Gamble provided content for Chapter 10.

# About the Author

David J. Teumim's background includes corporate security and web project management positions with Agere Systems and Lucent Technologies, along with 15 years of process, project, control, and safety work for Union Carbide Corp, British Oxygen, and AT&T.

His association with ISA began in early 2002 when he chaired ISA's first technical conference on Industrial Network Security in Philadelphia, PA, and taught the first ISA seminar on this subject.

Since 2004, his firm, Teumim Technical, LLC, has provided industry outreach for three U.S. Department of Energy National SCADA Test Bed projects, consulting for Sandia National Laboratories. More recently, he has chaired an American Public Transportation Association's Working Group on Control and Communications Security.

Teumim holds a master's degree in chemical engineering and is certified as a Certified Information System Security Professional (CISSP). He resides in Allentown, PA.

# Table of Contents

# Preface

So much has happened since the first edition of *Industrial Network Security* was published in 1995. This area has gone "mainstream" in terms of public awareness of the importance of Industrial Networks to our critical infrastructure and the threat to them from hackers, cyberspies, and cyberterrorists.

For instance, the story "America's Growing Risk: Cyber Attack" is featured on the cover of the April 2009 *Popular Mechanics*. And one of the lead stories on the front page of the 8 April 2009 edition of *The Wall Street Journal* was "Electricity Grid in U.S. Penetrated By Spies." The story talked about how foreign powers had mapped the U.S. electrical grid and left behind some rogue programs that could be activated remotely to disrupt the grid.

The "Big R," Regulation, has reared its head in the electric power industry. The NERC-CIP control system cybersecurity standards for electric power generation and transmission entities are now mandated by the U.S. government.

Commercial-off-the-shelf (COTS) hardware and software, as described in Chapter 3, continues its move into Industrial Networks as legacy equipment is phased out. And other sectors, such as passenger rail, described through the writer's eyes in the new Chapter 9, are coming up to speed on Industrial Network Security as COTS become commonplace in that sector control systems.

Consistent with the first edition, an effort has been made to keep this book introductory and easy-to-read. As with the first edition, this edition is intended for the technical layman, manager, or automation engineer without a cybersecurity background. New cyber incidents and updated information have been added to the chapters without changing the original format.

# 1.0
# Industrial Network Security

## 1.1   What Are Industrial Networks?

To define industrial network security, one first has to define industrial networks. For the purposes of this book, industrial networks are the instrumentation, control, and automation networks that exist within three industrial domains:

- *Chemical Processing* – The industrial networks in this domain are control systems that operate equipment in chemical plants, refineries, and other industries that involve continuous and batch processing, such as food and beverage, pharmaceutical, pulp and paper, and so on. Using terms from ANSI/ISA-84.00.01-2004 Part 1[1], industrial networks include the Basic Process Control System (BPCS) and the Safety Instrumented Systems (SIS) that provide safety backup.

- *Utilities* – These industrial networks serve distribution systems spread out over large geographic areas to provide essential services, such as water, wastewater, electric power, and natural gas, to the public and industry. Utility grids are usually monitored and controlled by Supervisory Control And Data Acquisition (SCADA) systems.

- *Discrete Manufacturing* – Industrial networks that serve plants that fabricate discrete objects ranging from autos to zippers.

The term Industrial Automation and Control Systems (IACS) is used by ISA in its committee name and in the recently issued standards and technical report series from the ISA99 Industrial Automation and Control Systems Security standards and technical committee (also, simply ISA99). This term is closely allied with the term *Industrial Networks*.

The standard, ANSI/ISA-99.00.01-2007-*Security for Industrial Automation and Control Systems, Part 1* [2], defines the term Industrial Automation and Control Systems to include "control systems used in manufacturing and processing plants and facilities, building environmental control systems, geographically dispersed operations such as utilities (i.e., electricity, gas, and water), pipelines and petroleum production and distribution facilities, and other industries and applications such as transportation networks, that use automated or remotely controlled or monitored assets." This standard will be referred to as "ISA-99 Part 1" in the book.

The technical report ANSI/ISA-TR99.00.01-2007 *Security Technologies for Industrial Automation and Control Systems* [3] succeeds the 2004 version of the document referenced in the first edition of this book. This report will be referred to as "ISA-99 TR1." Note: At the time of this writing, Part 2 of the ISA-99 standard has just been approved. Part 2 is titled *Security for Industrial Automation and Control Systems: Establishing an Industrial Automation and Control Systems Security Program* [4].

## 1.2    What Is Industrial Network Security?

When we speak of industrial network security, we are referring to the rapidly expanding field that is concerned with how to keep industrial networks secure, and, by implication, how to keep the people, processes, and equipment that depend on them secure. *Secure* means free from harm or potential harm, whether it be physical or cyber damage to the industrial network components themselves, or the resultant disruption or damage to things that depend on the correct functioning of industrial networks to meet production, quality, and safety criteria.

Harm to industrial networks and to the related people, processes, or equipment might be through the following:

- *Malicious Acts* – Deliberate acts to disrupt service or to cause incorrect functioning of industrial networks. These might range from a "denial-of-service" attack against a Human-Machine Interface (HMI) server to the deliberate downloading of a modified ladder logic program to a PLC (Programmable Logic Controller).
- *Accidental Events* – These may be anything from a "fat-fingered" employee hitting the wrong key and crashing a server to a power line surge.

When we think of industrial networks and computer-controlled equipment, we usually think of what ISA99 documents call "electronic security," but we should also include some aspects of two other branches of security: *physical security* and *personnel security*. These other two branches of security will be addressed in Chapter 2.

To illustrate the distinction, let's say we have a disgruntled employee who vents his anger in a chemical plant and:

1. turns a virus loose on the computer workstation that runs the HMI software, allowing the virus to spread through the industrial network;

2. takes a pipe wrench and breaks a liquid level sight glass on a storage tank, causing the liquid to leak out on the floor; and

3. pries open the door to a SIS system controller box and disables the overpressure shutdown by installing jumpers between isolated conductors and bypassing the audible alarms.

By our definition, acts 1 and 3 fall within our definition of industrial network security. Act 2 is deliberate sabotage, but it is physical sabotage of a mechanical indicating instrument, not of an industrial network. Act 3 involves some physical actions, such as breaking the lock and installing jumpers, but the jumpers then alter the electrical flow within an industrial network, a SIS system.

We acknowledge and stress the importance of physical protection of industrial network components, and also the personnel security that applies to the operators of these networks. However, physical and personnel security protective measures have been around for a long time, and information about these protective measures is readily available elsewhere. Chapter 2 introduces physical and personnel security as part of the entire security picture; however, the majority of this book covers the electronic security of industrial networks.

The ISA99 committee also acknowledges that these other branches of security, such as physical and personnel security, are necessary but similarly states that its standards are mainly concerned with the "electronic security" of industrial automation and control systems.

## 1.3   The Big Picture: Critical Infrastructure Protection

It is best to introduce the subject of Critical Infrastructure Protection from a historical perspective. In 1996, President Clinton issued PDD63 (Presidential Decision Directive 63) on Critical Infrastructure Protection[5], declaring that the United States had critical infrastructure that is vital to the functioning of the nation and must be protected. PDD63 identified eight critical infrastructure sectors, including these infrastructures using industrial networks:

- Gas and Oil Storage & Delivery
- Water Supply Systems
- Electrical Energy

Along with these three were also government operations, banking and finance, transportation, telecommunications, and emergency services.

In February 2003, President Bush released *The National Strategy to Secure Cyberspace*[(6)]. In it, some additional critical sectors were listed that use industrial networks, including:

- Chemical Industry
- Defense Industrial Base
- Food Production

Figure 1-1 shows how those original and additional critical infrastructure sectors map to the three industrial domains–chemical processing, utilities and discrete manufacturing–we described in Section 1.1 as using industrial networks.

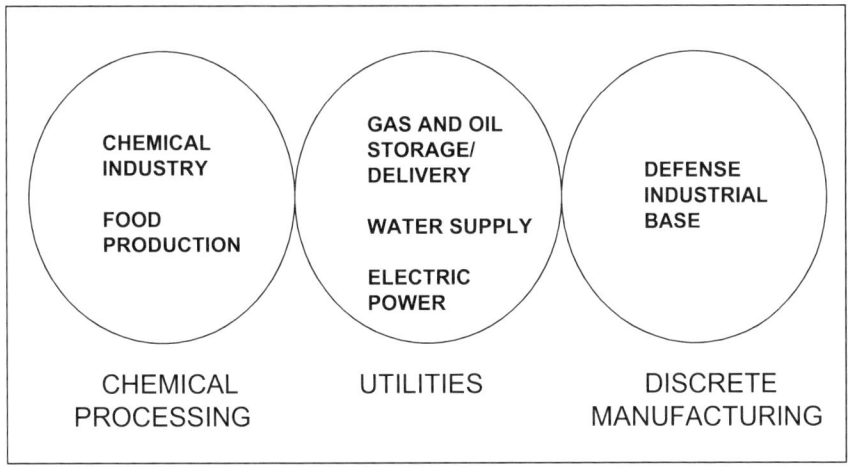

**Figure 1-1. Industrial Domain vs. National Critical Infrastructure Areas Using Industrial Networks**

The list of critical infrastructure sectors has continued to evolve since February 2003, with the federal government adding "critical manufacturing" to the list in 2008.

A glance at history shows how much the critical infrastructure sectors depend on each other–take one critical sector away and others may come tumbling down like dominoes. The Northeast Blackout of August 2003 showed how a failure of one sector may cascade to others. When the power went out in Cleveland, the water supply pumps in that city also shut down, since they ran on electricity. Similarly, the transportation sector in New York was affected when traffic lights ceased functioning and gas stations couldn't pump gas, since both were electrically operated.

What conclusions can we draw from this discussion of critical infrastructure?

We can conclude that securing industrial networks in our three domains of interest is a prerequisite for securing critical infrastructure at the national level. And this is true for all industrialized nations. In fact, the more automated and computer-dependent a nation's critical infrastructure is, the more it depends on developing and applying industrial network security to ensure its functioning in a new age of worldwide terrorism.

## 1.4   The Challenge: "Open and Secure"

Let's look at what has happened in the field of industrial networks in the last 12 years or so.

- *COTS.* Proprietary systems have given way to commercial off-the-shelf (COTS) hardware and software in industrial networks. Now we see everything from Microsoft Windows® to different flavors of Linux and Unix for operating systems, along with Ethernet, TCP/IP, and wireless protocols for networks.
- *Connectivity.* Once COTS hardware, software, and network components are used in industrial networks, the next logical thing is to connect the industrial networks and the business networks so the formerly incompatible systems can communicate. The business systems are invariably hooked up to the Internet.

- *Web, Web Services, and Wireless.* Recent developments include the ability to access a Web server in every intelligent electronic device and a browser on every engineer's office desktop to monitor equipment operations. And wireless LANs (Local Area Networks) offer the convenience of connecting devices without having to install expensive cabling within the plant.

All these developments have opened up our systems, but the question is, "Can we be both open and secure?" Being open and secure is the "Holy Grail" of our new industrial network security discipline. We want to keep the overwhelming business advantages of having open systems, yet secure our systems enough to ensure that our plants and utility grids don't become ready targets for cyber attack.

## 1.5   Who's Working on What?

For all practical purposes, the field of industrial network security began in the late 1990s. The September 11th attacks greatly accelerated the pace of activity. Since then, a bewildering variety of organizations with stakes in securing industrial networks have geared up to work on various aspects of the problem.

The organizations working on industrial network security may be divided into categories:

- *Government Organizations.* In the U.S., government agencies active in industrial network security include the National Cyber Security Division (NCSD) of the Department of Homeland Security (DHS), organizations within the Department of Energy (DoE), the DoE National Laboratories (e.g., Sandia, Pacific Northwest, and Idaho National), the Department of Commerce National Institute of Standards and Technology (NIST), the Federal Energy Regulation Commission (FERC), and the General Accounting Office (GAO). Each organization has some stake in protecting the industrial networks that make up portions of the nation's critical infrastructure. Some organizations, such as FERC, now have regulatory authority, as will be discussed in 1.6.

- *In the international arena,* government organizations like Canada's Office of Critical Infrastructure Protection and Emergency Preparedness (OCIPEP) and Britain's Centre for Protection of National Infrastructure (CPNI) play a similar role in protecting their nation's critical infrastructure.

- *Nonprofit Organizations.* These range from international professional and technical societies spanning industrial sectors, like ISA, to U.S.-based industry sector-specific groups like the North American Electric Reliability Corporation (NERC) for electric power and the American Water Works Association (AWWA) for the water utilities. Included among the nonprofits are schools and universities that have courses, seminars, and research and development programs in industrial network security.

- *For-Profit Entities.* The various corporations that are the vendors and users of industrial networks are key in determining whether industrial network security procedures and equipment are developed, commercialized, purchased, and used successfully.

Within the organizational categories listed above are two organizations that deal with industrial network security, working at the international level across the three areas of chemical processing, utilities, and discrete manufacturing.

These organizations are:

- ISA, through technical and standards committees like ISA99, Manufacturing and Control Systems Security.

- IEC (International Electrotechnical Commission), including Committee 65 for work on the IEC 62443 Network and System Security Standards.

These organizations work across industrial areas and, therefore, manufacturing sectors. For instance, we previously mentioned the ISA-99 series of standards and technical reports that define the breadth of "Industrial Automation and Control Systems" as "applied in the broadest possible sense, encompassing all types of manufacturing and process facilities and systems in all industries in every area of manufacturing."

## 1.6   Federal Regulatory Authority

Recently, two federal groups have been given regulatory authority over industrial network security in the public and private sector. The Federal Energy Regulatory Commission has been given the authority to regulate the cybersecurity of the transmission grid, and it has exercised that authority by making the NERC CIP (North American Reliability Corp. Critical Infrastructure Protection) Consensus Industry Standards into official federal regulations with enforcement penalties. The Department of Homeland Security with their CFAT (Chemical Facility Anti-terrorism) Regulations on the chemical industry, are mostly concerned with physical security but have a cybersecurity section. Other departments of the federal government regulating other critical infrastructure sectors may well get into the act in the future.

## References

1.  ANSI/ISA-84.00.01-2004 Part 1 Functional Safety: Safety Instrumented Systems for the Process Industry Sector – Part 1. ISA, 2004.

2.  ANSI/ISA-99.00.01-2007 *Security for Industrial Automation and Control Systems, Part 1*. ISA, 2007.

3.  ANSI/ISA-TR99.00.01-2007 *Security Technologies for Industrial Automation and Control Systems*. ISA, 2007.

4. ANSI/ISA -99.00.01-2007 *Security for Industrial Automation and Control Systems: Establishing an Industrial Automation and Control Systems Security Program, Part 2.* ISA, 2007.

5. The White House. Presidential Decision Directive 63. *Protecting America's Critical Infrastructure.* May 22, 1998. Retrieved 11/11/2004 from: http://www.fas.org/irp/offdocs/pdd/ pdd-63.htm.

6. The White House. *National Strategy to Secure Cyberspace.* February 2003. Retrieved 11/11/2004 from: http://www.whitehouse.gov/ pcipb/cyberspace_strategy.pdf.

# 2.0
# A Security Backgrounder

## 2.1 Physical, Cyber, and Personnel Security

When considering security for business and industry, security practitioners have traditionally divided themselves into three areas of specialization. We describe these three areas with the aid of two terms used frequently in security:

- *Insiders.* The people who belong in your facility, including employees and invited contractors, visitors, or delivery and service people.

- *Outsiders.* People who don't belong in your facility, whether they enter physically or electronically. This category covers everyone from vendors through hardened criminals! Uninvited outsiders in your facility are intruders and are guilty of trespassing, at the least.

Keeping these terms in mind, and as mentioned in Chapter 1, the three traditional areas of security are:

- *Physical Security.* Guards, gates, locks and keys, and other ways to keep outsiders from becoming intruders and insiders from going where they don't belong. This is the oldest and most established branch of security and claims the highest percentage of security professionals.

- *Personnel Security.* Practitioners here are usually occupied with these questions: "Are the outsiders I'm about to bring into my plant trustworthy?" and "May I continue to place trust in my insiders?" This area of the security profession covers everything from criminal background checks on new employees and contractors to investigation of security violations by employees and periodic background rechecks of existing insiders.

- *Cybersecurity.* This category covers prevention, detection, and mitigation of accidental or malicious acts on or involving computers and networks. The area now known as business or IT cybersecurity has its roots in the financial and intelligence communities of the 1960s and 70s.

Industrial network security is primarily IT cybersecurity adapted to industrial networks, but includes important elements of physical and personnel security as well. For instance, does it make a difference if your valuable process recipes, kept as trade secrets on your control network, are taken by industrial spies who:

- hack into your industrial network through the corporate firewall and business network and then download and sell them? (a cybersecurity incident), or

- pull up in a van disguised as legitimate messengers from your computer tape backup storage firm and get an unwitting employee to hand over your freshly made backup tapes containing the same trade secrets (a personnel security incident), or

- break into your plant late at night, cleverly bypassing the burglar alarm, and walk out with the hard drives from your control servers containing the recipes (a physical security incident)?

The net effect is the same in all three incidents–your secrets are gone! In fact, an industrial spy may purposely "case the joint" and choose an attack plan based on where your defenses are weakest.

Successful prevention of industrial network attacks involves getting knowledgeable specialists from all three areas of security to sit around the table and discuss possible attacks and means to prevent them. Brainstorming techniques may be used, with no type of attack dismissed as "too wild an idea" to consider.

For example, before the Sept. 11, 2001 attacks, the philosophy driving airline security was "hijackers want to live." Wouldn't it have been valuable to question that assumption in the years leading up to September 11 and say, "But suppose the hijackers want to die? What could or would they do then?"

In this writer's experience in the corporate security world, I would sit at the lunch table listening to corporate security investigators tell stories of active investigations. Many of their stories were bizarre, such as employees using their corporate credit cards to pay for anything from expensive parts for their own motorcycles to thousands of dollars in elective surgery! Any rational employee would say, "Don't do that, you'll get caught!" Did these employees think about consequences before they went ahead with their plans? Maybe, but the consequences didn't deter them from going ahead anyway.

Let's see if we can brainstorm a scenario of factory sabotage. For example, the successful sabotage of a factory conveyor system might (1) involve an unscrupulous salesman from a rival conveyor company who has a criminal record (personnel security). (2) He strays into the production area while left unattended after visiting the engineering department (physical security). (3) There, he downloads a modified ladder logic program from his laptop to the conveyor machinery PLC (cybersecurity). That causes the conveyor to mysteriously malfunction the next day, making a purchase of his company's rival conveyor system more likely the next time he pays a sales call!

Analyses of security incidents usually reveal a chain of events that led up to the actual criminal activity. If security measures, whether they involve physical, personnel, or cybersecurity activity, can be introduced to prevent, detect, and respond to the chain of activities at any point, there is a good chance the final criminal activity can be prevented.

In the conveyor system example, where might security have been introduced to interrupt the chain of events leading up to the conveyor sabotage? Would the outcome have been different if:

- the rival conveyor company had done a criminal background check in the hiring phase and discovered that the salesman had a criminal record; or

- the factory he was visiting had a "company escort required" physical security policy, preventing the salesman from wandering into the production area alone; or

- the factory had active network security measures that prevented the salesman from entering the PLC network and downloading a modified ladder logic program?

If any of these physical, personnel, or cybersecurity measures had been in force, the final event in the chain, the conveyor's mysterious malfunction, might have been prevented.

## 2.2   Risk Assessment and IT Cybersecurity

Risk assessment is the process by which you and your management team make educated decisions about what could harm your business (threats), how likely they are to occur (likelihood), what harm they would do (consequences), and, if the risk is excessive, what to do to lower the risk (countermeasures).

Let's say you are the owner of a large factory making widgets in a Midwestern state, which happens to be in "Tornado Alley." Your plant building and attached business office building are as shown in Figure 2-1:

For instance, for risks to the office building and its contents, such as the business computer systems, we can illustrate what one type of risk assessment—a quantitative risk assessment—looks like. In this example

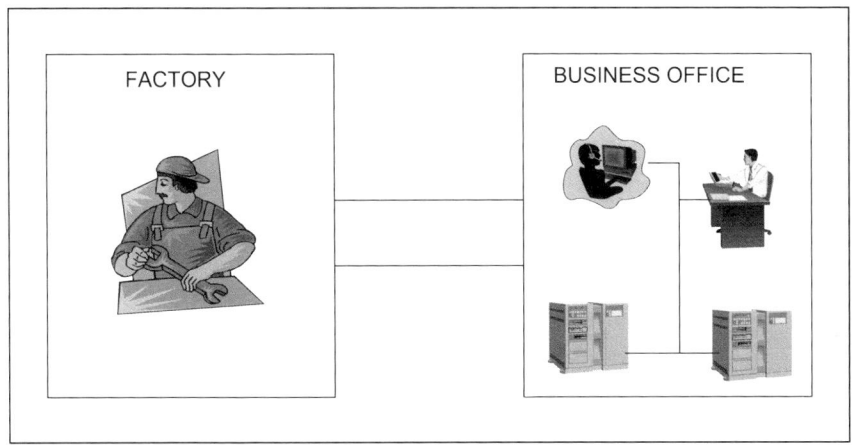

**Figure 2-1. Widget Enterprises, Inc.**

we will consider one physical and one cyber threat to the office building and its computer system, per Figure 2-2.

The first, a mild-to-moderate tornado, represents a physical risk to the office building and its contents. Let's say the likelihood of a mild-to-moderate (known as category F0 to F2) tornado hitting the office building is once every 20 years (a fairly dangerous neighborhood!). The figure assumes the consequence of the threat or average damage to the asset (office building) is $5 million. Therefore, the annual risk from mild-to-moderate tornado damage is:

$$1 \text{ event}/20 \text{ years} \times \$5 \text{ million/event} =$$
$$0.05 \times 5 =$$
$$\$0.25 \text{ million/year at risk from this type of tornado.}$$

Now we have a measure of annual risk in terms of dollars. We can compare it with the very different risk of, let's say, a particular type of cyber attack by an industrial spy who seeks to download your carefully guarded database of best customers and what they typically order from you.

| THREAT | LIKELIHOOD (in number of events per year) | ASSET | CONSEQUENCE | CONSEQUENCE (in $) | RISK (in $ per year) | COUNTER MEASURE |
|---|---|---|---|---|---|---|
| PHYSICAL (TORNADO) | One event/ 20 years, or .05/year | Office building | Damage to building + contents | $5 million per event | 0.05/year X $5 million = $0.25 million | Reinforced concrete construction to limit damage |
| CYBER (INDUSTRIAL SPY) | One event/ three years or 0.33/year | Customer database | Lost sales | $10 million per theft | 0.33/year X $10 million = $3.3 million | Strengthen business office cyber defenses |

**Figure 2-2. Office Building – Physical and Cyber Risk Assessment**

Once we enter the cyber realm, doing a quantitative risk assessment raises a problem: unlike weather damage or a physical security issue like robbery, there are not a lot of historical statistics to draw from to get likelihood numbers. But some data on the frequency of industrial spying of all types does exist, with on-average loss by different size companies and industries. This data, coupled with loss data from your factory, might enable you to come up with a reasonable estimate so you could continue being quantitative (as opposed to qualitative, which is the alternative. We will focus on qualitative risk assessment in an upcoming section).

Let's estimate the likelihood of this event at one cyber-theft (threat) every three years, and the sales you would lose as a result of this information being given to your competitors (consequence) at $10 million. Then, from this type of cyber event:

1 event/three years × $10 million = $3.3 million/year at risk.

Here is the power of a quantitative risk assessment. For the first time, we can compare the cost of physical damage to cyber damage in terms that top management will understand–dollars. Based on this risk assessment, we may conclude that the monetary risk of an industrial spy cyber attack is greater than the monetary risk of a tornado. In later chapters, we will see how countermeasures or preventive remedies, such as reinforced construction to limit tornado damage, can be evaluated against calculated risk to see if they are worthwhile.

Keep in mind that our risk analysis has been simplified. Usually, more terms enter into a risk analysis, and, as mentioned, getting good numbers or ranges of numbers for a quantitative cyber risk assessment may be difficult.

The following people will have a lot of interest in the office building risk assessment we just made:

- The business owner, the CEO, and the general managers
- The Physical Security Manager and the Facilities Manager (who may be the same individual)
- The Chief Information Officer (CIO) and the part of the CIO's organization responsible for business systems cybersecurity (perhaps an IT cybersecurity manager)

Let's draw an organization chart (see Figure 2-3) to represent a simplified management structure for a stand-alone factory. (Note that in a modern multi-plant manufacturing corporation, numerous "dotted line" relationships would exist between corporate and plant management.)

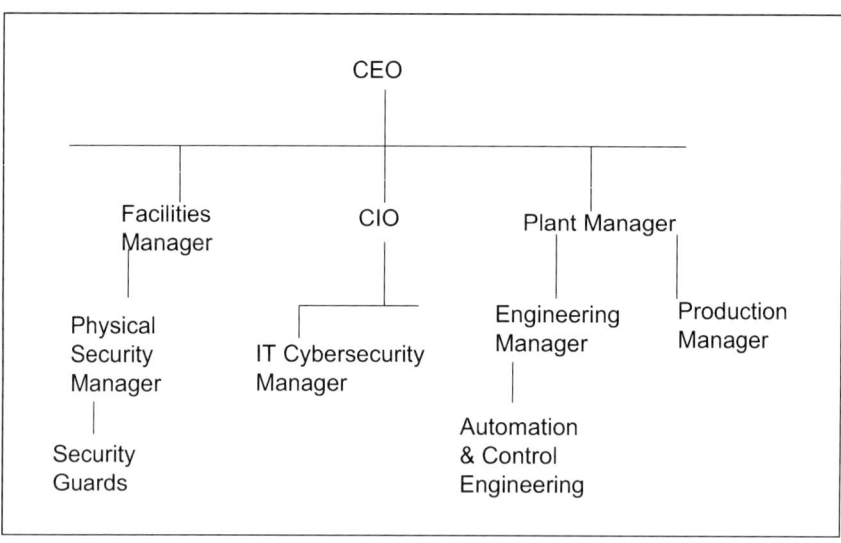

**Figure 2-3. Organization Chart**

The IT cybersecurity manager, who reports to the CIO, is responsible for the corporate firewalls and Intranet and Internet access, and might have these IT security issues to deal with:

- *Web.* Downloading of pornography or illegal content by employees.

- *Email.* Viruses coming in; spam.

- *Remote access.* Allowing authorized users to connect via modem pool or virtual private network, and keeping unauthorized people and hackers out.

- *Unlicensed software.* Keeping employees from using unpaid-for or unapproved software.

To address these problems and a host of other IT security issues, the IT cybersecurity manager draws on the field of business or commercial cybersecurity. This field, termed "computer and network security" in prior times, includes the following:

- *IT security technology.* Firewalls, antivirus programs, and audit and security diagnostic programs and tools.

- *Trained personnel.* Specially trained computer security practitioners, holding certifications such as Certified Information System Security Professional (CISSP) or Certified Information Systems Auditor (CISA) and trained in the IT security body of knowledge.

- *IT security policies, processes, and procedures.* Published cybersecurity guidelines and recommendations from various commercial cybersecurity organizations.

In short, a "body of knowledge" is readily available for this area, whether we call it IT, commercial, or business cybersecurity.

## 2.3 Risk Assessment for the Plant

Now that we've covered the business office building, let's take a look at our widget production factory building (Figure 2-4):

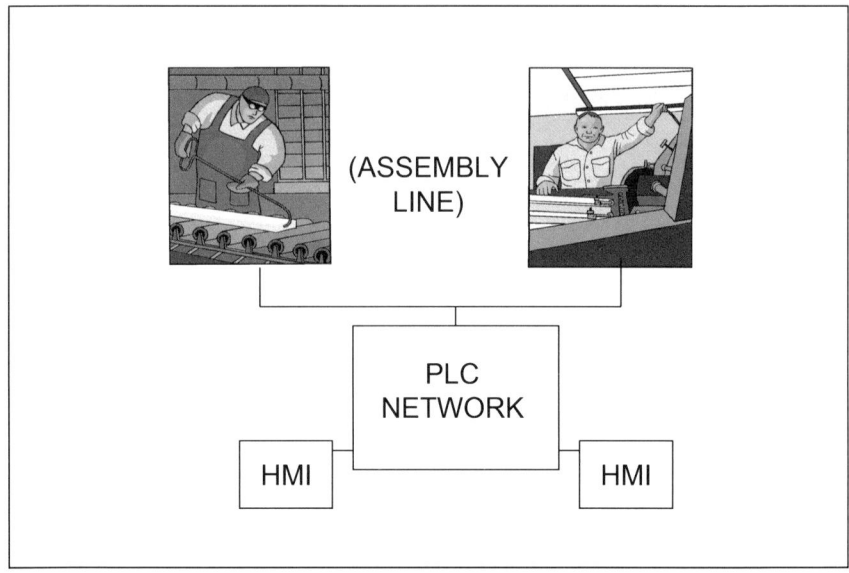

**Figure 2-4. Inside the Factory Building**

Here, we see the type of industrial network we would expect to see in discrete manufacturing, with PLCs, HMIs, etc.

This time, let's illustrate a risk assessment more appropriate to a plant scenario, where we may not have access to realistic numbers or estimates for the likelihood of a physical or cyber attack. In a qualitative risk assessment, relativity rankings substitute for absolute numbers or estimates of likelihood and consequences. The output is a prioritized list of risks, showing which are more substantial.

Figures 2-5 and 2-6 give the procedure for a qualitative assessment and the resulting risk matrix. We are evaluating two scenarios here. The first—a physical attack—is a sabotage of the assembly line by a disgruntled employee with hand tools. The second is a cyber attack to sabotage the PLC network that runs the assembly line.

1. Risk team lists a number of security scenarios to consider:
   a) Physical sabotage of assembly line machinery by disgruntled employee with hand tools
   b) Cyber sabotage of PLC network controlling assembly line

2. Rank order the a) Likelihood and b) Consequence of these scenarios on a relative scale. For instance:

| Likelihood | Consequence to Plant |
|---|---|
| 1) Not Likely | 1) Not Critical |
| 2) Somewhat Likely | 2) Somewhat Critical |
| 3) Likely | 3) Critical |
| 4) Very Likely | 4) Very Critical |

3. Let's say the risk team rates scenario a) (physical sabotage) as 2) somewhat likely, and 3) of critical consequence to plant, and it rates scenario b) (cyber sabotage) as 3) likely and 3) of critical consequence to plant

**Figure 2-5. Qualitative Risk Assessment Example**

As a result of the risk assessment process shown in these figures, the risk assessment team concludes that scenario (b), the cyber attack, is more threatening than scenario (a), the physical attack.

# 2.4  Who's Responsible for Industrial Network Security?

Now we come to the question, "Who's responsible for the (1) physical security and (2) cybersecurity of the industrial network?"

Let's look at a possible list of candidates. Within the CIO organization, there might be an IT cybersecurity manager, per the organizational chart in Figure 2-3. Within the factory organization any or all the following managers and technical people might be involved:

- Plant Manager
- Production Manager

1. The risk team plots the two scenarios on a QUALITATIVE RISK MATRIX below

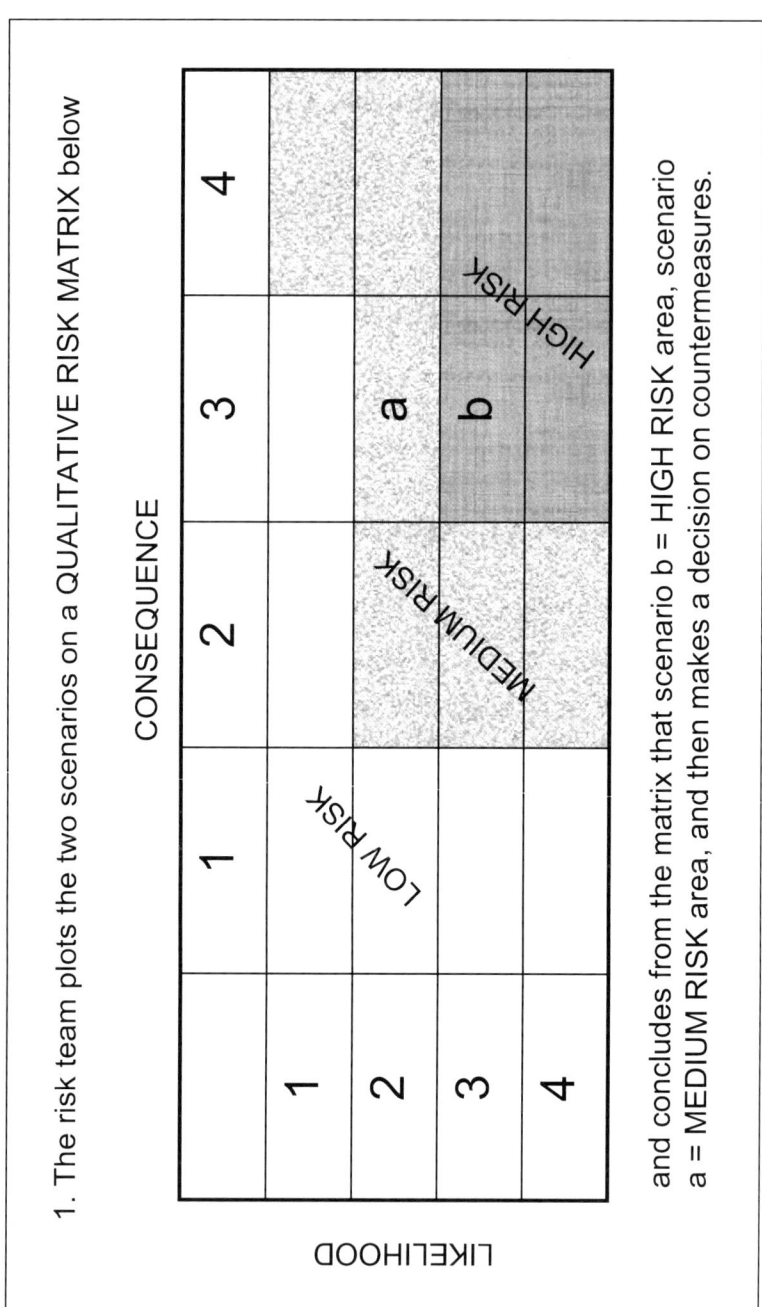

and concludes from the matrix that scenario b = HIGH RISK area, scenario a = MEDIUM RISK area, and then makes a decision on countermeasures.

**Figure 2-6. Qualitative Risk Matrix**

- Engineering Manager

- Automation and Control Manager

- Automation Engineer, Technician, and Plant Operator

- Facilities Manager

- Physical Security Manager

So who do the CEO and upper management usually think is responsible for industrial network physical and cybersecurity? For the physical security of the industrial network, it may be argued that whoever is in charge of plant physical security, such as the Facilities or Physical Security Manager, has this responsibility. (Although the plant security guards are usually guarding the plant entrances, far away from the production area of the factory, this might theoretically cover the disgruntled employee attacking the PLC network with a pipe wrench!)

But, in many conference discussions the author has participated in, the usual answer is that if the CEO and top management realize that industrial network cybersecurity is a legitimate concern at all, they think the CIO and the IT cybersecurity manager have this area covered. (And they usually point to the corporate firewall, corporate cybersecurity policies, and the gamut of IT security controls to prove it.)

But if we then go to the CIO organization and ask the IT cybersecurity managers how well they are covering this "newly assigned" area of industrial network security, the typical answer might be they are totally unfamiliar with control systems: "Engineering and Production handle that."

As mentioned, the field of industrial network security really began in the late 1990s and then accelerated following the September 11 attacks. Since September 11, a lot of progress has been made in this field by the many organizations listed in Section 1.5 of this book. However, in contrast to IT cybersecurity, the field is still young and there is only a limited amount of knowledge and experience to draw upon. And unless a corporation has had the foresight to specifically designate an individual or a group, or its entire Automation and Control Engineering staff, to handle this very specialized area of industrial network security, the real answer to who is responsible for industrial network security is "no one!"

Unlike the commercial computing profession, which has included cybersecurity as a legitimate area of study and practice for many years, the automation and controls area has not traditionally had much contact with any area of security, especially cybersecurity. Security, whether physical, personnel, or cyber, is just not in the curriculum of the vast majority of engineering and technical schools. It is slowly making its way into the curriculum in some universities in the form of individual courses and seminars, but is certainly not in the mainstream yet.

Many manufacturing corporations that decided to build an organization or entity to handle industrial network security have formed a cross-disciplinary task force, committee, or permanent group, consisting of people and/or knowledge and experience from the following plant organizations:

- Automation and Controls Engineering, Production, and Maintenance
- IT Cybersecurity
- Safety (especially in a hazardous workplace, such as a chemical plant or refinery)
- Physical Security (facilities)
- Human Resources (for personnel security matters)

Only when industrial network security is included as part of an overall security effort will the proper resources, leverage, and empowerment be available to do the job well. Although grassroots efforts by control engineers to secure their industrial networks are well-intentioned and commendable, they will seldom be enough to do the job. Just as with safety, the first step starts with ownership and commitment by upper management.

But, as mentioned, top management may not recognize a clear need for an effort in this area. A business case for industrial network security may have to be made and presented. The following section gives some tips on how to do this.

## 2.5  Tips for Making the Business Case to Upper Management

1.  Don't use cyber "tech-talk" to sell top management on industrial network security. Instead, use a language they understand—risks, consequences, and the cost of reducing the risk versus the cost of doing nothing. As much as possible, try to put consequences in dollar terms.

2.  Don't use the "sky-is-falling" approach and concentrate only on the worst-case scenario. That gets old fast. Instead, add up the consequences of inaction—whether it be a threat to safety, lost trade secrets, downtime, etc. Even better, try to include all possible consequences in an itemized scenario.

3.  Do be very specific. If production downtime is a consequence, how many days of downtime? What will the cost be? What will be the cost of getting production going again, of cleaning up a virus from the industrial network, for instance?

4.  Do realize that you can't protect everything from every threat. Countermeasures to reduce the risk usually cost money. And the necessity of spending the money to pay for these countermeasures will have to be sold to management.

(This is a process called risk management, which we will cover later in this book.)

5. Do use publicly documented cases in which industry was hit by cyber attacks. Some well-documented cases of cyber attacks are described in Chapter 4. Then describe what the consequences would be if a similar attack hit your plant or industry.

## 2.6   Making the Business Case with Data

Here is an example of how a business case was made for a significant IT cybersecurity investment[1].

A Texas University medical center cybersecurity manager calculated the cost of spam to his organization at $1 per spam message, and the cost of recovering from the Nimbda outbreak in 2001 at $1 million. On the basis of these numbers, he successfully justified to the chief financial officer the purchase of spam filtering and enterprise antivirus software and showed how the countermeasures would more than pay for themselves. The business case was made with hard business data from his organization, in dollars.

A similar approach might be used to argue for industrial network security. Let's say you are a control engineer using COTS software on your industrial network and have had the good fortune never to have been hit by a virus or worm. If your control network is part of a large multinational corporation, chances are that some portion of the IT network in your corporation was hit. And it probably has downtime and network recovery figures that you can use for your estimates, as well as horror stories.

By asking the question "If this attack had happened to our industrial network(s), what would the result be in, say, X number of servers down, Y days of lost production, Z days to clean up and recover?" You might make a convincing case that, since major virus/worm attacks happen at least several times a year, your company might avoid the inevitable loss by installing countermeasures such as firewalls, antivirus software, or other products.

## References

1.  Violino, B. "Texas University Calculates Financial Benefits of its Spam, Virus Defense." InternetWeek.com article. October 29, 2003. Retrieved 11/11/2004 from: http://www.internetweek.com/showArticle.jhtml?articleID=15600902.

# 3.0
# COTS and Connectivity

## 3.1 Use of COTS and Open Systems

Commercial-off-the-shelf (COTS) describes the movement of business and commercial computer and networking hardware and software into the industrial network area, displacing proprietary devices and applications. This trend started 10 to 15 years ago and includes the following:

- Operating systems. Microsoft Windows NT®, Windows 2000®, and Windows XP® are being used in industrial networks. In the Unix world, flavors of Unix including Sun Microsystems' Solaris®, IBM's AIX®, and Hewlett-Packard's HP-UX®, to name a few, have also moved into industry. Most recently, the Linux world has entered industrial networks.

- Database software, such as Microsoft SQL Server® and Oracle® databases.

- Hardware, including Windows® PCs, workstations, and servers, and Unix workstations and servers.

- Networking products such as Ethernet switches, routers, and cabling.

- Networking protocols for TCP/IP-based LANs, using protocols such as HTTP, SNMP, FTP, etc.

- Development languages, including C++, Microsoft Visual Basic.NET®, Microsoft C#®, Sun's Java®, etc.

- Object Linking and Embedding for Process Control (OPC).

- Internet, with standard or custom browsers as process interfaces to web servers in IEDs (Intelligent Electronic Devices).

- Wireless LANs using the IEEE 802.11 protocol.

## 3.2   Connectivity

Once COTS is used in industrial networks, the business side demands, "Now that you have opened it up, connect it so we can talk."

Connectivity is desired:

- between the corporate business network and the industrial network,
- for remote access to the industrial network from outside the corporate firewall, and
- to vendors, customers, and other business partners from the industrial network.

## 3.3   What You Get that You Didn't Bargain For

The movement to COTS and connectivity gives you a multitude of business advantages, such as:

- Standardization
- Compatibility with business systems
- Much lower purchase cost
- Familiar interfaces
- Less training time and effort

With these advantages, you also get some "baggage" to contend with:

1. Forced updates to software are much more frequent than with the original proprietary systems.
2. There are millions of extra lines of software code for a multitude of features, many not wanted or needed in industrial applications.
3. The industrial world is not the business driver for COTS.

4. Numerous software-related quality and security issues exist, in part the result of the drive by vendors to get new software out the door quickly.

5. There is a continual need to install patches for software security and proper functionality.

These drawbacks are seldom realized up front, when the systems are purchased.

The business concept called "total cost of ownership" enables you to realistically evaluate these systems by adding the cost of maintenance, updates, patching, etc., to the up-front purchase or licensing cost over the life of the installed system. When doing a total cost of ownership analysis, these life-cycle costs should be included in the analysis. This concept is discussed in Reference 1.

It is apparent that some of the economic benefits of moving to COTS and connecting up are negated by some of the drawbacks. For instance, how many proprietary industrial network software programs have ever been hit by a computer virus or worm?

Remediation of attack by a virus or worm is a hidden cost of using COTS, which will not show up during purchase but which should be included in a total cost of ownership analysis. If antivirus software is purchased to prevent these cyber attacks, the cost of installing and maintaining this software should also be included in the total cost of ownership analysis.

# References

1. Emigh, Jacqueline, "Total Cost of Ownership." Computer-world.com article. December 20, 1999. Retrieved 11/11/2004 from: http://www.computerworld.com/hardwaretopics/ hardware/story/ 0,10801,42717,00.html.

# 4.0
# Cybersecurity in a Nutshell

## 4.1  Security Is a Process

Security is very similar to safety in that it is a continual process rather than an endpoint. A control network that is secure today may be insecure tomorrow, because hackers are always thinking up new attacks.

Securing industrial networks involves technology, but technology is only one ingredient of the final mix. Successful industrial network security is a carefully composed mixture of the following:

- Educated and aware users

- Appropriate organizational structure

- Security strategy matched to the organization structure

- Policies and procedures that work

- Audit and measurement programs

- Security technology appropriate to the above mix, at a level of sophistication understood by those who use it

## 4.2  Basic Principles and Definitions

We can carry over some basic principles of commercial computer and network security to the industrial network space. The first is called the AIC triad. AIC stands for Availability, Integrity, and Confidentiality. Figure 4-1 shows these principles as the points of a triangle:

Let's start with availability. For industrial networks, availability means the network is fully operational and available to users and other machinery and processes when needed. If the system is not operating,

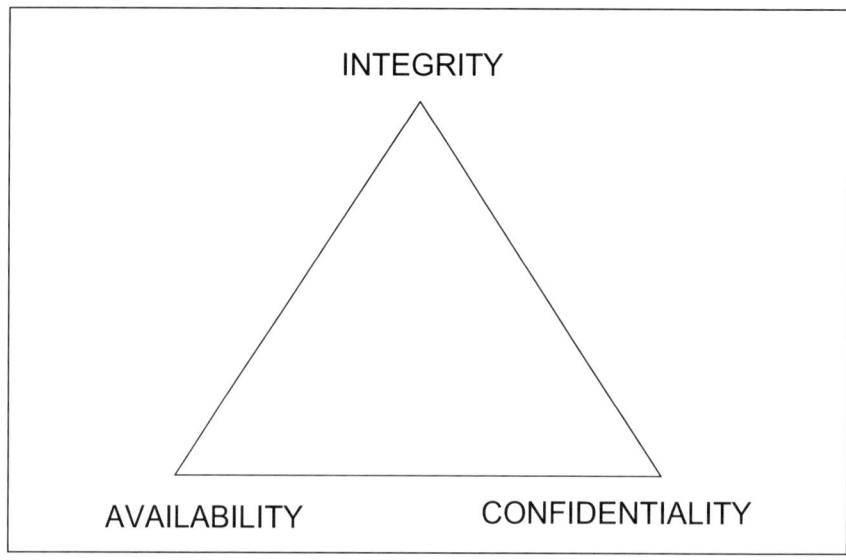

**Figure 4-1. The AIC Triad**

or not operating correctly for any reason when it is needed, this property is not satisfied. It could be unavailable for many reasons, such as the following:

- An unintentional user error crashed the system.

- The system has a computer virus or was just hacked by an insider or outsider.

- A power failure has occurred, and the backup generator isn't supplying enough power.

- The computer room just burned to the ground.

---

### Case History 1: Lack of Availability

The Omega Engineering logic bomb: Omega Engineering is an instrument and control vendor in New Jersey that suffered heavy losses in May 2000 when it fired a disgruntled computer systems administrator[1]. Before he left the building, the employee planted a "logic bomb," which, when activated, erased Omega's production software programs. He also stole the company's software backup tapes as "insurance"!

It took Omega Engineering months to get back into production after this incident. The company suffered heavy financial losses, while their competitors gained ground on them.

The next AIC factor is integrity. Integrity in computer security may be defined from two angles: the integrity of the data, and the integrity of the computer hardware and software itself.

Integrity of data means that there should be no inadvertent or malicious modification of data while it is stored or being processed on a system.

Let's apply this concept to a SCADA system for a gas pipeline. If a remote pressure sensor on the pipeline reads 1000 psig (process data), and that value is faithfully transmitted to the central gas control room and shows up as 1000 psig on the main control panel, we have data integrity. If the value shows up as 2000 psig or 500 psig, we have a process data integrity problem!

Hardware/software system integrity implies that the hardware and software versions and configuration are correct at any given time, and only authorized changes or updates have been made.

For instance, hardware/software integrity is flawed if an HMI application was tested only with a previous release of an operating system, and the operating system software is upgraded or patched without proper compatibility testing and change authorization.

The third AIC component is confidentiality—the ability to keep information on a computer system secret. It should be accessible only to people authorized to receive and view and modify that information, and no one else.

For instance, a chemical or pharmaceutical corporation has recipes, formulas, and production methods it wants to keep away from competitors and to prevent the information from becoming public knowledge. The company has gone to great lengths to develop or acquire this information.

---

**Case History 2: Theft of Trade Secrets**

A case involving Lucent Technologies[2] illustrates the significance of confidentiality in computer security. In 2001, two Chinese nationals were indicted for stealing proprietary telecommunications computer code while working at Bell Labs in Murray Hill, New Jersey. They were first noticed when their employer observed portions of the proprietary computer code being emailed from the company's network. They were successfully convicted in one of the first cases prosecuted under the 1996 Federal Economic Espionage Act protecting trade secrets.

---

# 4.3  Basic Principles: Identification, Authentication, and Authorization

In addition to the AIC triad, three other important definitions are important in classic computer and network security: identification, authentication, and authorization.

Identification answers the question, "Who am I?" If I log on to my computer as user DJT, that tells the computer I am David J. Teumim, a legitimate user listed in the password file.

But how does the computer distinguish me from an imposter posing as me?

Authentication requires that you "prove it" by reinforcing your identity, using one or more of three possible authentication factors:

- Something you know (a password)

- Something you have (a hardware token or key)

- Something you are (a biometric, like your voiceprint or finger-print)

Using more than one authentication factor increases security.

For instance, several chemical companies use "two-factor authentication" to grant employees remote access to plant computers from their homes. The hardware token (something you have) displays a unique number that changes every minute according to a random pattern. When the remote user logs in, he or she enters the number on the token, along with a four-digit fixed PIN number (something you know). The random number entered by the user must match the pre-synchronized random number on the company's central security administration server. Only then is the user granted remote access rights.

Authorization deals with what your access privileges are, once you have successfully logged on to the protected system. Which system features may you use? Which system programs or files may you view, modify, delete, etc.?

For instance, in the control room of a petroleum refinery, control room operators may have access to functions required for normal operation, but only control engineers may be authorized to perform other functions, like changing HMI programming.

## 4.4   More Cyber Attack Case Histories

This section describes some control system attacks that have been documented in the press.

## Case History 3: SCADA Attack

This incident is a classic in industrial network security, the first publicly documented cyber attack on a control system, in this case, a wastewater treatment SCADA system in Australia.

In this incident[3], a 49-year-old man who had worked for the supplier that installed a computerized SCADA system for the municipal wastewater works was convicted of a cyber attack on the municipality's sewage system. The attack sent millions of gallons of raw sewage spilling into local parks and rivers in Queensland, Australia, causing considerable damage. The convicted man was caught with radio equipment and other computer apparatus used to hack into the SCADA network in his car.

## Case History 4: Computer Worm in a Nuclear Plant Control System

In August 2003, the Nuclear Regulatory Commission (NRC) issued an information alert to all nuclear plant operators about a situation that occurred earlier in 2003 at the Davis-Besse nuclear power plant in Ohio[4], which was infiltrated by the Slammer worm. In a scenario all too familiar to IT cybersecurity experts, the worm entered the plant by a roundabout route. A T1 communications line that led to a network to which the company's corporate business network was connected became the conduit for the worm to reach and crash the Safety Parameter Display System (SPDS). The SPDS system is an industrial network that displays the status of critical reactor safety monitoring sensors such as core temperature, coolant status, etc. Fortunately, the plant was off line, and a backup analog system could be used while the digital system was out.

### Case History 5: Computer Worms Infect Auto Manufacturing Plant

In August, 2005, thirteen DaimlerChrysler auto manufacturing plants were knocked offline for an hour by two Internet worms, idling 50,000 workers, while infected Windows 2000® systems were patched[5]. The Zotob and PnP worms infected systems integral to the manufacturing process.

Could the incidents described in Case Histories 3, 4, and 5 have been prevented? Chances are excellent that with a sufficiently advanced and well-thought-out industrial network security program, they could have been. However, even in the best-planned schemes, there is no foolproof program to ensure you will never have a security incident. If prevention fails and you do have an incident, the goal of industrial network security is to detect the threat and mitigate the damage as quickly and efficiently as possible.

# 4.5 Risk Assessment and Risk Management Revisited

Let's return to our discussion of risk assessment, begun in Chapter 2.

Suppose we have an industrial network controlling our factory's assembly line. The assembly line machinery can be attacked physically, by a disgruntled employee, or by an outside hacker who can get into the system by several means.

We introduced these terms in Chapter 2:

- Asset (What you have that you want to protect.)
- Threat (The person or event that can cause harm.)
- Consequence (The harm that can be caused.)

- Likelihood (How often the threat is expected to cause harm over a certain time.)

- Risk (Consequences expected over a certain time.)

- Countermeasures (Ways to reduce risk.)

Let's now look at cyber threats in more detail, and add another term to our risk assessment model: vulnerability.

## 4.6  Cyber Threats

Military, law enforcement, and IT cybersecurity experts typically break down the category of threats further, in what is known as "threat analysis."

We can introduce the following terms and concepts:

- Adversary (Who is he, she, or it? Is it a single person, an organization, or a terrorist group?)

- Intent (What motivates this person or organization? Anger? Revenge? Money?)

- Ability (How capable is your adversary? Able to write custom scripts for cyber attack? Or merely capable of downloading scripts that others write, and then running them?)

- Target (What is their immediate goal? Their ultimate goal?)

Let's construct a simple chart, a threat matrix, to describe these concepts for several threat agents (see Figure 4-2).

## 4.7  Vulnerabilities

A vulnerability is a "chink in your armor," an inviting spot or situation where an attack by an adversary is likely to succeed. For instance, if a

| ADVERSARY | INTENT | ABILITY | TARGET |
|---|---|---|---|
| SCRIPT KIDDIE | MISCHIEF, BRAGGING | LOW | ANY SYSTEM |
| INDUSTRIAL SPY | THEFT | VERY HIGH | TRADE SECRETS |
| DISGRUNTLED EMPLOYEE | REVENGE | VARIABLE | INDUSTRIAL NETWORK ? |

**Figure 4-2. A Threat Matrix**

burglar tries your locked front door and then goes around to the back door and finds it unlocked, the unlocked back door is a vulnerability.

In industrial network security, a vulnerability is a place where a cyber attacker can bypass whatever built-in defenses an application, network, or operating system has in order to gain privileges that would normally be unavailable. This enables the attacker to insert actions and commands, or even become the all-powerful system administrator on an operating system like Windows, or acquire "root" privileges on a Unix box.

Using COTS hardware, software, and networking in industrial networks brings into the controls world the same vulnerabilities that plague the Internet and the business computing world. COTS software vulnerabilities are due to the following:

- *Complexity.* Operating systems and application software have millions of lines of code. One figure quoted in the literature says there is an average of one software bug per 100 lines of code. Some fraction of these bugs will be security vulnerabilities. (Figure out how many software bugs are in a 40 million line program!)

- *Inadequate Quality Assurance.* Software manufacturers do not always catch these quality and security flaws before they go out the door as production code. They may think it sufficient to use software customers as "quality testers" and have them report bugs to be corrected in the next software revision.

- *Speed to Market.* Competition and concentration on numerous new features lead to rapid-fire releases of new software versions.

- *Lack of Seller Liability.* The majority of commercial software licenses do not hold the seller responsible for any damage to your systems from software that does not function properly. (Contrast that with the liability for manufacturers of cars, household appliances, or airplanes. If these products cause injury or economic damage, a rash of lawsuits usually follows, sometimes involving punitive damages.)

- *Lack of Security-Based Development Tools and Languages.* The standard software development languages, such as C, C++, and Visual Basic, were not composed with security in mind. Adding security features was frequently an assigned or unassigned task left up to the programmer, who is under development time pressure. This situation is beginning to change, as there are now seminars, books, and some software tools to help the developer write more secure software.

Let's look next at the most common COTS software flaw affecting security—the buffer overflow.

## 4.8   A Common COTS Vulnerability: The Buffer Overflow

Buffer overflows cause an estimated 40 percent of the exploitable software flaws in the COTS software environment. Sad to say, they have been around for more than 20 years. We know how to fix this flaw, but the discipline to eliminate buffer overflows has not permeated very far into COTS software development.

In programming languages, such as the C language, when you run a function (which is somewhat like a subroutine) from the main program, the memory area devoted to your function will contain a "stack," or buffer area. The stack contains things such as the values you are calling the function with, and the local variables you will be using in the function. At the end of the allotted buffer space for the function is a "return address" that tells the computer what line in the main program to return to after it has finished running the function.

Suppose, in the C language, you want to ask the user for input via the keyboard as a task for your function. Say you want to ask the user for his or her "last name," and you figure it should be no more than 20 characters long.

You would assign a variable like "Lastname" to hold 20 characters maximum. But the C language lacks an inherent mechanism for preventing a malicious user from putting in too many characters when typing input, and the computer will accept those extra characters and allocate those extra and unexpected characters to "Lastname" in the buffer.

A clever hacker can craft a very long string of characters, followed by a short, very carefully constructed command that overwrites the original return address sitting in memory at the end of the allocated buffer space. The new return address tells the computer to return to a place in the hacker's code, not to the legitimate address that was in the original program. This overruns the buffer when the input is given.

If the hacker is clever enough to craft the right commands in that illegitimate string, he or she can insert commands that will give "root" privileges on a Unix box or administrator privileges on a Windows operating system when overflowing certain programs. Essentially, the hacker now "owns" the system, with one buffer overflow command. Not a bad achievement for a hacker who can craft the right string!

The clever original hacker who discovered the buffer overflow string may then publish the technique to a hacker website or bulletin board for other, less-experienced "script kiddies" to use.

As we have seen, despite the fact that buffer overflows have been known for about more than 20 years, and programming techniques have been developed to fix them, progress on eliminating them has been slow. New code comes out every day with buffer overflow vulnerabilities just waiting to be discovered. Once they are discovered in published software code (let's hope by someone on the security side of the fence and not a hacker), the only hope is for the software supplier to issue a code fix or "patch" for systems administrators to apply before a new cyber attack takes advantage of the vulnerability.

# 4.9    Attacker Tools and Techniques

Let's look at some of the tools and techniques our adversaries use:

- *Viruses*. Viruses have been around since the advent of the PC. They spread by infecting new host computers with their code (which can be carried on a USB flash drive or CD), by a program, or a by macro for a spreadsheet or word processing program. A virus can spread by email if it contains an executable attachment that can be opened.

- *Worms*. A worm contains self-replicating code that may spread through a network like a LAN or the Internet. A worm spreads copies of itself and does not need host software to spread.

- *Trojan Horse*. This is a program that seems to do something beneficial with one part of the code, while a hidden part of the code does something malicious. An example of a Trojan Horse would be a screensaver that also emails a copy of the confidential data files on your computer to a competitor!

- *Logic Bomb.* This software program lies dormant on a computer hard drive until it is activated by a trigger, such as a certain date or event. Then it activates and causes malicious activity.

- *Denial-of-Service Attack.* This kind of attack, usually network-based, overwhelms a server with a flurry of false requests for connection or service, causing the server to lock up or crash.

- *Botnets.* Botnets are networks of infected computers available to do the bidding of "bot herders" who rent out their hundreds or thousand of compromised computers for hacking or coordinated denial-of-service attacks.

The hacking community spreads its know-how and wares through a variety of outlets:

- *Hacking websites.* Thousands of websites across the Internet offer advice and code on everything from stealing phone service to breaking into wireless networks. Such sites may even offer downloadable "point-and-click" hacking tools for the novice.

- *Books and CDs.* At most local computer shows, you can find inexpensive CDs loaded with hackers' tools and "exploit code."

- *Chat Rooms and Bulletin Boards.* Many hackers will brag about their techniques and offer to share them in online chat rooms like Internet Relay Connection (IRC).

# 4.10 Anatomy of the Slammer Worm

Now that we've seen how our adversaries (disgruntled employees, industrial spies, and hackers) can get their hands on tools (viruses, worms, network scripts that exploit vulnerabilities in COTS code), let's take a look at a 2003 worm called Slammer that caused the nuclear plant safety display monitoring system shutdown described in Section 4.4.

The Slammer worm caused havoc, bringing the entire Internet to a crawl in just 15 minutes. The attack started with a single data packet, a User Datagram Protocol (UDP) packet of 376 bytes total (much smaller than previous worms such as Code Red, at 4 KB, or Nimbda, at 60 KB). It targeted UDP port 1434, the port that Microsoft SQL (Structured Query Language) Server database software listens in on. Once received, Slammer overflowed the buffer with specialized code that spilled past the 128 bytes of memory reserved for the input. It then had machine-language code that caused the machine to overwrite its own code and reprogram itself to send out a flurry of new 376-byte UDP packets to Internet IP (Internet Protocol) addresses it calculated using a random number generator. The timing was such that the worm could double the number of infected hosts every 8.5 seconds, bringing the Internet, and corporate LANs connected to it, to a crawl as the available bandwidth was used up.

As the previous section indicates, the Slammer worm clogged up internal bandwidth at the Davis-Besse nuclear plant industrial network. It also caused considerable damage elsewhere. A 911 call center in Washington State that used the SQL Server database was effectively shut down. Emergency dispatchers had to resort to a cumbersome manual procedure to make do until the system could be brought back up.

A synopsis of how the Slammer worm spread is shown in Figure 4-3.

# 4.11 Who's Guarding Whom?

One final observation will add a bit of irony to round out our discussion of COTS software vulnerabilities. Let's assume we have a software-based firewall to protect an internal LAN that we connect up to the Internet. We need this firewall to prevent Internet-based attacks like

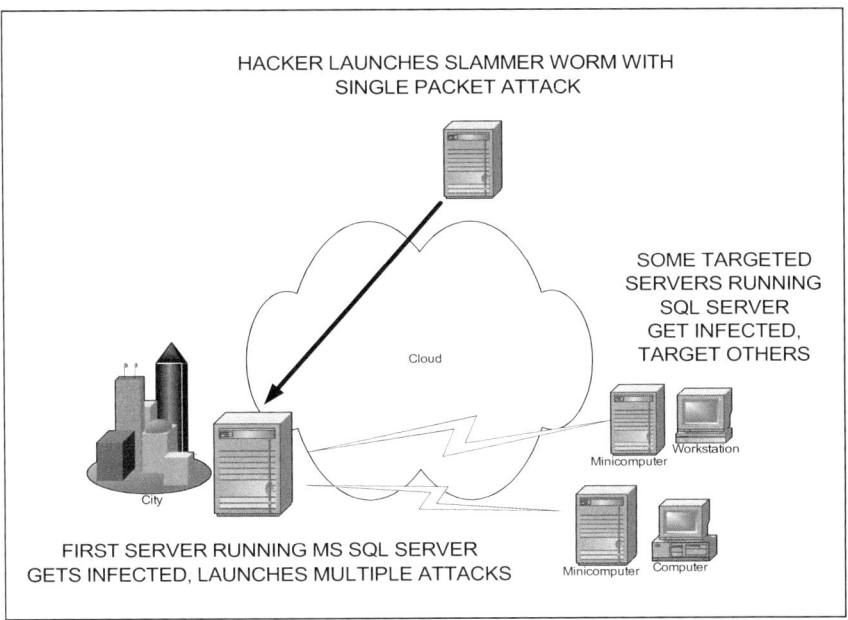

HACKER LAUNCHES SLAMMER WORM WITH
SINGLE PACKET ATTACK

SOME TARGETED
SERVERS RUNNING
SQL SERVER
GET INFECTED,
TARGET OTHERS

Cloud

Workstation

Minicomputer

City

FIRST SERVER RUNNING MS SQL SERVER
GETS INFECTED, LAUNCHES MULTIPLE ATTACKS

Minicomputer    Computer

**Figure 4-3. How the Slammer Worm Operates**

worms, and other network attacks, from reaching our internal hosts because we know the software on our internal hosts on our LAN might be susceptible to (for example) buffer overflow attacks.

So our software-based firewall is "guarding the gate" against cyber attacks that exploit buffer overflow vulnerabilities. This gives us a warm feeling of security until we find out that our firewall code itself may contain buffer overflow vulnerabilities! (Note: Security researchers regularly find and publish information about software bugs and vulnerabilities [including buffer overflow attacks] within security software, such as software-based firewalls and antivirus software).

Once these vulnerabilities are found and published, the only alternative for security-conscious systems administrators is to patch and patch again. There is an area of expertise called "Patch Management" that is now applicable to industrial networks to address how, when, where software patches should be applied. Within industrial networks, a patch management program assumes a very important role because critical infrastructure is involved.

# References

1. Ulsch, M. *Security Strategies for E-companies.* Infosecuritymag.com column "EC Does It," July 2000. Retrieved 11/11/2004 from: http://infosecuritymag.techtarget.com/articles/july00/columns2_ec_doesit.shtml

2. United States Department of Justice *"Former Lucent Employees and Co-conspirator Indicted in Theft of Lucent Trade Secrets."* Cybercrime.gov press release, May 31, 2001. Retrieved 11/11/2004 from: http://www.cybercrime.gov/ComTriadIndict.htm

3. Schneier, B. *The Risks of Cyberterrorism*, Crimeresearch.org article taken from TheMezz.com, June 19, 2003. Retrieved 11/11/2004 from: http://216.239.39.104/custom?q=cache:uJQl__6DhAUJ:www.crime-research.org/news/2003/06/Mess1901.html+Schneier&hl=en&ie=UTF-8

4. Poulsen, K. *Slammer Worm Crashed Ohio Nuke Plant Network*, Securityfocus.com article, August 19, 2003. Retrieved 11/11/2004 from: http://www.securityfocus.com/news/6767

5. Roberts, P. F.*Zotob, PnP Worms Slam 13 DaimlerChrysler Plants*, August 18, 2005. Retrieved 8/8/2009 from http://www.eweek.com/c/a/Security/Zotob-PnP-Worms-Slam-13-DaimlerChrysler-Plants/

# 5.0
# Countermeasures

## 5.1 Balancing the Risk Equation with Countermeasures

In our discussion on risk assessment thus far, we have been adding terms to our list of risk assessment factors from previous chapters to arrive at the list below:

- Asset
- Threat
- Consequence
- Likelihood
- Vulnerability
- Risk
- Countermeasures

Let's take a look at the interrelationships among the first six terms in Figure 5-1. Then, in Figure 5-2, let's see how countermeasures fit in.

Now that we have illustrated the relationships between the risk terms with and without countermeasures, let's see, on a more practical level, how countermeasures might be introduced into our quantitative and qualitative risk assessment examples from Chapter 2.

## 5.2 The Effect of Countermeasure Use

Figure 2-2 (Chapter 2, Section 2.2) showed a simple risk assessment illustration for the office building connected to the widget factory. In it,

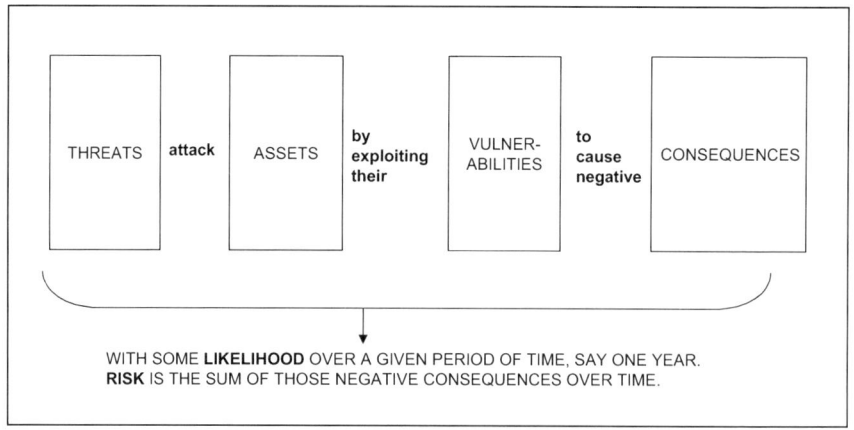

**Figure 5-1. Risk Assessment Before Countermeasures**

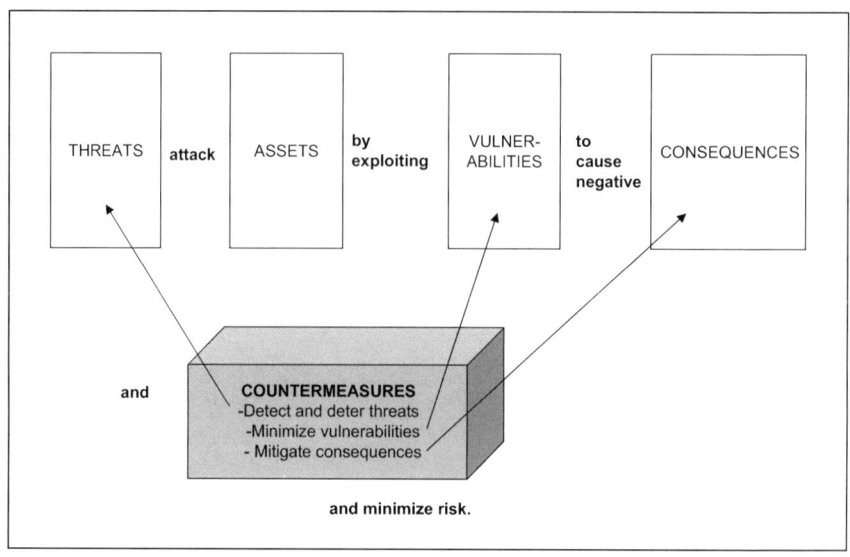

**Figure 5-2. Risk Assessment Adding Countermeasures**

we see that the risk, or expected loss per year from a mild-to-moderate tornado striking the office building, is $.25 million, or $250,000 per year.

Now suppose we want to introduce a countermeasure to reduce the expected loss per year. We can compute the cost of reinforcing the office building structure and spread that cost out over the same number of years as our risk assessment time frame figure, 20 years. (Note that this is a rather simplistic analysis in terms of the reality of financing building improvements.)

Let's say reinforcing the walls and roof to prevent tornado damage will cost $1 million, and we do this today. The risk evaluation for the reinforced building covers the next 20 years. So $1 million/20 years = $.05 million or $50,000 cost per year for 20 years.

Now let's calculate the reduction in expected loss per year by reinforcing the building. Our risk was $.25 million, or $250,000 per year, so spending $50,000 per year on countermeasures will reduce risk by $250,000. (Note: in practice, countermeasures are rarely 100 percent effective. A certain amount of damage risk per year, termed residual risk, would probably exist despite your best efforts at building reinforcement.)

Not bad—we have spent $50,000 per year to save $250,000 in risk. Neglecting residual risk, our net saving by risk reduction is:

$250,000 saved/year – $50,000 spent on countermeasures = $200,000/year. It still looks like a good deal!

Figure 5-3 shows the risk assessment for the building after adding tornado countermeasures.

Now suppose instead we spend $5 million to reinforce the building and evaluate that over 20 years. Would this be a good decision? Well, $5 million/20 years = $0.25 million/year. We would spend $250,000 on countermeasures to save $250,000 on annual risk. Our net savings in estimated loss per year would be zero!

| THREAT | ASSET | CONSE-QUENCE (in $$) | RISK (in $/YEAR) | COUNTER-MEASURE | COUNTER-MEASURE COST | YEARLY RISK REDUCTION |
|--------|-------|---------------------|------------------|-----------------|----------------------|----------------------|
| PHYSICAL (TORNADO) | Office building | $5 million per event | 0.05/year X $5 million = $0.25 million | Reinforced concrete construction To limit damage | $1M/20 years = $50,000/year | $200,000 per year |
| CYBER (INDUSTRIAL SPY) | Customer database | $10 million Per theft | 0.33/year X $10 million = $3.3 million | Strengthen business office cyber defenses | ---- | ----- |

**Figure 5-3. Office Building – Physical and Cyber Risk Assessment**

We can see that we are in a powerful position if we are fortunate enough to have historical weather damage data to draw from to support a quantitative risk assessment. We can calculate when a countermeasure will pay for itself and at what point it does not make economic sense.

The same type of analysis can be made for our industrial cyber spy scenario in Figure 2-2. However, we should remember that our risk numbers and the effect of countermeasures will be more estimated and, therefore, more open to variability.

Let's turn to how we can evaluate the effect of countermeasures in a qualitative risk assessment. With a qualitative risk assessment, we do not deal directly in dollars. Instead, we determine which risks are greater, then prioritize the spending of our resources on countermeasures.

Let's go back to the factory risk assessment from Chapter 2, Section 2.3, and the qualitative risk assessment process and matrix shown in Figures 2-5 and 2-6. As Figure 2-6 shows, scenario (a) (physical attack) produces a "medium" risk rating, and scenario (b) (cyber attack on the PLC network) produces a "high" risk rating.

If we can introduce countermeasures to decrease the likelihood of a cyber attack, then we might be able to move scenario (b) from the "high" risk zone to the "medium" risk zone, alongside scenario (a). We might do this by better isolating the PLC network from the rest of the company and the outside, or by decreasing cyber vulnerabilities, or by mitigating the effects of a successful cyber attack with a quicker or more complete disaster recovery program.

Discussion might focus on which approach(es) would lower risk level most, what countermeasure(s) to use, how effective each would be, and so on. The cost of each alternative countermeasure might be estimated, for example, along with how effective it would be in reducing total risk.

So when we evaluate the effect of countermeasures in reducing total risk in a qualitative risk assessment, we are really going through a process analogous to our quantitative example.

A risk management step normally follows the risk assessment step, with the assessment team weighing the results of the risk assessment step.

There are three possible risk management decisions the team can make once they know what the risks are:

- Accept the risk
- Minimize or eliminate the risk
- Transfer the risk

Accepting the risk means essentially to do nothing. The enterprise chooses to live with the risk and accept the consequences should it happen.

Minimizing or eliminating the risk means countermeasures will be evaluated and applied. (And the residual risk, left over after countermeasures are applied, will be accepted).

The third alternative transfers the risk to another party, such as an insurance company. For instance, the enterprise will pay an insurance premium for protection from loss of sales in the event of a sabotage attack.

The remainder of this book deals with constructing an industrial network cyber defense. In other words, we are assuming the second risk management option and focus on minimizing or eliminating risk, if possible, by using countermeasures.

## 5.3 Creating an Industrial Network Cyber Defense

After we have done a qualitative risk assessment, we may decide to go with the second risk management option and focus on minimizing or eliminating risk, if possible, by taking countermeasures. How do we go about deciding on what countermeasures are appropriate for industrial networks in our chemical plants, utility grids, and factories? Chapters 6–8 of this book deal with constructing an industrial network cyber defense, but we'll look at it briefly here.

Figure 5-4 summarizes the contents of Chapters 6 through 8. It shows the "Countermeasures" block from Figure 5-2, separated into physical and personnel security countermeasures, together with the topics of Chapters 6–8 as components of an overall cyber defense.

As shown in Figure 5-4, a good industrial network defense contains the following:

- Design and Planning
- Technology
- People, Policies, and Assurance
- Physical and Personnel Security Countermeasures and Support

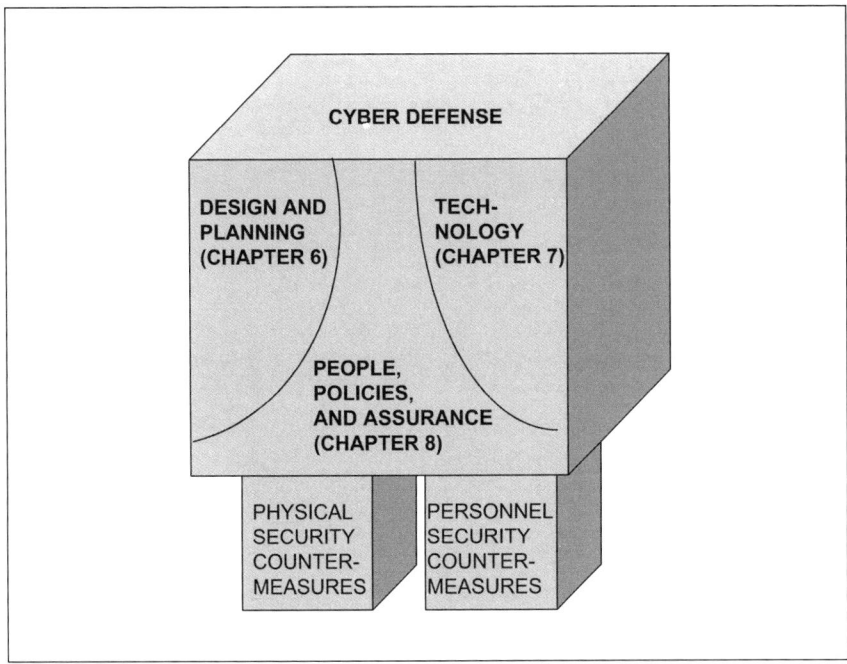

**Figure 5-4. Countermeasure Components**

Countermeasures may act in a variety of ways, as the face of the countermeasures block of Figure 5-2 shows. Countermeasures may act to:

- deter and detect the threat (as a barking watchdog on the premises would detect and deter a burglar),

- minimize a vulnerability (as bars on a window would make forced entry more difficult), and

- mitigate the consequences (as effective disaster recovery plan gets a hacked server up and running again).

# 6.0
# Cyberdefense Part I – Design and Planning

## 6.1 Defense in Layers

The principle of defense in layers is that one relies on many different overlapping layers to prevent a worst-case scenario. If one layer fails, the next is there to take over, and so on.

To understand how this concept may be applied to industrial network security, let's first look at the way the concept is applied in a common chemical processing application that incorporates a Safety Instrumented System (SIS).

One simple polymerization process uses two hazardous chemicals, a monomer (chemical A) and a second reactant (chemical B), which may be an initiator or catalyst for the reaction. The reaction is exothermic, which means heat is released when the two chemicals are combined and brought up to reaction temperature.

Figure 6-1 shows an example of the simple polymerization reaction setup. In it, our monomer (chemical A) flows from a storage tank on the right through a control valve into the reactor, where it combines with chemical B, which flows from the storage tank on the left, through a control valve, and to the reactor. The process may be sequential (i.e., first the monomer is charged to the reactor, then chemical B is added slowly during the actual reaction step).

A well-known process safety hazard of polymerization is the possibility of a "thermal runaway," where the reaction heat builds up inside the

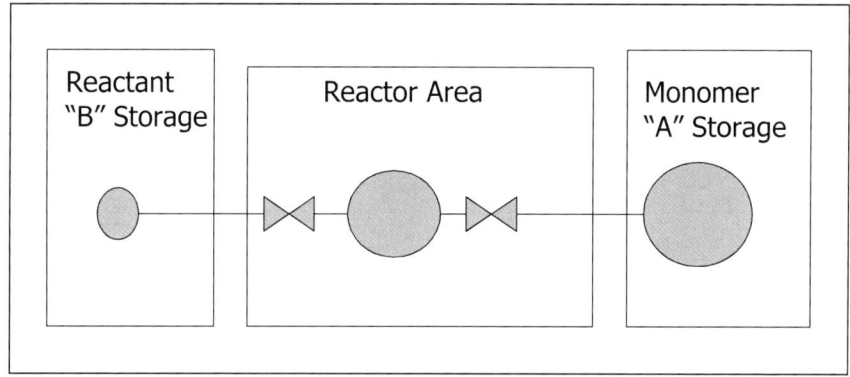

**Figure 6-1. Polymerization Plant Example**

reactor vessel, raising the temperature and pressure of the reaction mixture until it bursts the reactor vessel, leading to an explosion, fire, and hazardous fluid release into the surroundings. The process safety strategy is to keep the reaction under control by removing the heat that is generated, never letting it build up to the point where the reaction produces more heat than can be removed.

Reference (1) gives a case history of a polymerization reactor runaway and explosion that was investigated by the U.S. Chemical Safety and Hazard Investigation Board.

To counter the possibility of a thermal runaway, control systems safety design uses "layered defenses"[2]. Protection in layers forms the foundations of SIS design by such specifications as ANSI/ISA-84.00.01-2004, Functional Safety: Safety Instrumented Systems for the Process Industry Sector, and IEC 61508, Functional Safety of Electrical/Electronic/Programmable Electronic Safety-Related Systems. The system designer contains the hazards of this process by successive layers of control and mechanical systems protection, as shown in Figure 6-2[3]:

The layers of protection against a runaway reaction begin with the basic process control system (BPCS). If control of the process from the BPCS is lost and the reaction temperature and pressure go too high, then, in

```
┌─────────────────────────────────────────────────────────────┐
│                                                               │
│   SECONDARY CONTAINMENT, PLANT, COMMUNITY EMERGENCY RESPONSE  │
│                                                               │
│      EMERGENCY RELIEF DEVICES (RUPTURE DISKS) ACTIVATE        │
│                                                               │
│         SIS (SAFETY INSTRUMENTED SYSTEM) ACTIVATES            │
│                                                               │
│         HIGH TEMPERATURE/PRESSURE ALARMS SOUND – OPERATOR     │
│         MANUAL INTERVENTION                                   │
│                                                               │
│           BPCS (BASIC PROCESS CONTROL SYSTEM) TRIES TO CONTAIN│
│                                                               │
│             REACTION OVERHEATS, RUNAWAY STARTS                │
│                                                               │
└─────────────────────────────────────────────────────────────┘
```

**Figure 6-2. Layers of Protection Against a Runaway Reaction**

the next layer, alarms on excessive reaction temperature and pressure will sound, requiring manual action by operators to shut down the reaction process.

If these layers fail–the alarm malfunctions, the operators don't respond or respond incorrectly, etc.–then the next layer, the SIS, will take over. In our example, this might be done by shutting off the flow of reactant B and/or by providing emergency cooling.

The next layer is mechanical (for example, blowing the rupture disk to release the reaction contents). After that, additional layers might include a secondary containment system (dikes, etc.), and, finally, emergency response, first by the plant and then by the community.

These layers of protection should be as independent as possible, so the failure of one layer does not affect the performance of the next.

## A Security Example

Now let's say our polymerization takes place in a small chemical plant that has an office building located beside the control room as shown on the site layout in Figure 6-3. (In reality, the control room and office building should be located a safe distance from the reaction area and chemical storage.) Note that in the safety example, the hazard we were protecting against arose inside the reaction vessel, and our layers extended outward around it. In this security example, we are protecting from the outside in.

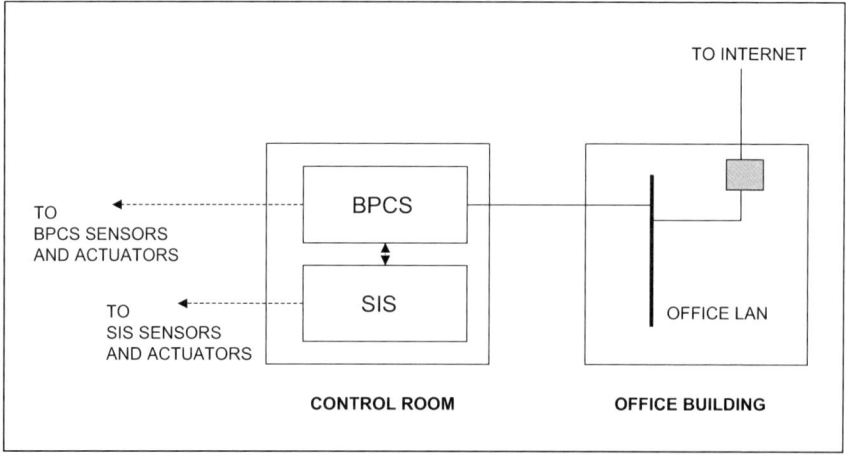

**Figure 6-3. Polymer Plant Site Layout**

Let's include the business and control networks in Figure 6-3. The business network will serve the office building, and the control room/chemical reactor area will have a Basic Process Control System (BPCS) network and a Safety Instrumented System (SIS).

Let's say our task is to protect the office network, the BPCS, and the SIS from a hacker who is bent on causing a runaway reaction by using the Internet to penetrate the chemical plant through the firewall. Above all, we want to protect the SIS, since it is a critical safety system. Next in importance to the process is the BPCS and, finally, the business system.

Drawing a series of concentric rings around first the SIS, then around the BPCS, and finally around the business network, as shown in Figure 6-4, will help us discuss defense in layers for security.

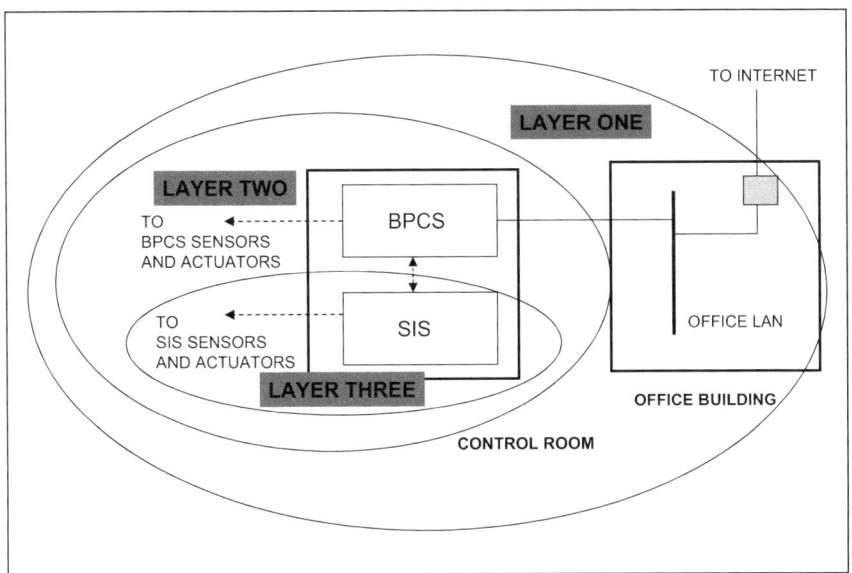

**Figure 6-4. Cyber Defense in Layers**

A cyber attacker would first have to penetrate the corporate firewall to get to the business network (Layer One). The next target would be the BPCS network (Layer Two), and finally the SIS (Layer Three). If only the business network and BPCS are compromised, the SIS and subsequent safety layers will act to prevent a runaway. If both the BPCS and the SIS are compromised, a runaway is more likely. It can now be prevented only by additional protection layers like operator action or mechanical safety devices such as rupture disks and secondary containment. If all else fails, the consequences would be dealt with through emergency response.

For a cybersecurity defense in layers to be effective, each layer should have its own defenses and not merely "sit by" passively. For instance, the business network might have an intrusion detection/protection system to detect and prevent cyber attacks from beyond the firewall.

However, suppose we attach an external modem to the BPCS network in Figure 6-4, so the process engineers can telecommute to the plant on weekends and holidays. What happens to our defense in layers model now? If an outside hacker, through war dialing and password guessing, can obtain entry to the BPCS in one step instead of having to hack in through the corporate firewall, he has effectively bypassed Layer One and is at Layer Two. (A war dialer is a computer program used to identify phone numbers that can connect with a modem.) Even worse, if there is a modem connection into Layer Three, perhaps to let the SIS vendor communicate with the SIS, the hacker might bypass both Layers One and Two to gain access. The hacker might commit hidden sabotage to Layer Three, perhaps by deactivating the SIS. This might not become obvious until the BPCS loses control of the reaction, and the SIS is needed to bring the reaction back into control.

This brings up another observation: Each layer of defense is effective only if there is no easy way to bypass the layer.

## 6.2   Access Control

Access control for industrial networks is the important area of determining and enforcing who (or what device or system) has access to the system assets, such as the HMI, the process control network, the controllers, servers, etc. And, if a person, device, or system is allowed to "touch" these system assets, access control specifies:

- What is their authorization level?
- What data or settings may they change, delete, add, etc.?
- How will this be controlled and enforced?

Along with cyber access control, the parallel area of physical access control will determine and enforce who can walk into the control room or other physical location where the industrial networks are located. To be truly effective, cyber and physical access control must act together.

So let's continue with our illustrative example of the small polymerization plant illustrated by Figures 6-1 through 6-4, and see how access control integrates with the "defense in layers" model.

Although it might not be typically thought of in this fashion for a defense in layers model, we might visualize Layer One in this example as having two regions:

1. A perimeter, or boundary
2. An interior area

It is easy to visualize these two Layer One regions in the office LAN in Figure 6-4. The corporate firewall separates the office LAN from the Internet. The firewall represents region 1 above, the perimeter or boundary, separating inside from outside. The office LAN, on the other hand, extending through the office building and interconnecting many different servers and workstations, is the interior area and represents region 2.

It is just as important to the success of the defense in layers model for the interior region, the office LAN, to be "hardened," that is, not to have obvious network or host vulnerabilities, as it is for the firewall to be correctly configured, monitored, and maintained. What happens within the office LAN is crucial to maintaining the effectiveness of the perimeter protection of the firewall. Both the perimeter and the interior of Layer One must act together.

For example, let's say the firewall is configured and operating perfectly. If an office worker receives a piece of malicious email containing an executable of a Trojan Horse, his or her machine may be "taken over" and used to launch attacks on the connecting networks. Some Trojans can even establish an outbound connection from the office LAN host that was taken over that goes out through the firewall to the hacker's server on the Internet. The outgoing traffic from the machine that has been taken over will look like an innocent web (http) connection initiated by that internal host.

For another illustration of the concept of defense in layers, let's now consider both physical and cyber access control of Layer Two. Physical access control would regulate who can come into the control room, which may have a locked door with only authorized employees having the key, for instance. Once inside the control room, an employee would need the proper cyber access, a correct login and password, to access BPCS control functions. Access control also includes authorization levels, which might allow control engineers to change process set points but not allow operators to perform the same actions.

It also would be desirable to have a third person in the loop, a control network administrator, who would assign and administer the logins, passwords, and authorization levels in step with personnel changes. In the following sections of this chapter, we will discuss different security aspects that, taken together, lead to the success of the defense in layers security strategy.

The above discussion, where we visualize each layer of protection as composed of a perimeter and an interior area, is formalized in the ISA-99 Part 1 standard as the "zone and conduit" method for Industrial Network Security.

The zone and conduit method becomes the tool for risk assessment and then risk management and reduction. The interior area comprising Layer One becomes the "zone," where risk level is uniform, and the corporate firewall connecting Layer One with the Internet becomes the "conduit." Readers are referred to ISA-99 Part 1[(4)] for further details.

# 6.3   Principle of Least Privilege

One concept we will borrow from IT cybersecurity for use in industrial network access control is called "the principle of least privilege," also known as "security by default." In theory, this principle is straightforward, but in practice, applying this principle is very difficult in a conventional plant control room with operators, supervisors, and engineers logging on to consoles using a typical system of user logins and passwords. If we were to apply the principle of least privilege to access control in a control room, we would do the following:

- Start by denying everything. Deny all access and authorization to everybody.
- After proper identification and authentication, grant access and authorization privileges (the ability to do authorized tasks) for only those minimum sets of functions each individual needs to do his or her job, and no more.
- Remove these access and authorization privileges promptly when the individual no longer needs them, such as after a new assignment or job rotation.

Many longtime employees in the process industries "accumulate" passwords—and therefore unneeded access and authorization privileges— as they rotate through various jobs. The principle of least privilege requires organizations to keep track of what access and authorization privileges an employee needs to perform present tasks, and to allow authorization for those functions only.

If an employee or contractor leaves or is terminated for cause, by far the most important access control action to perform is to remove all physical and cyber access and authorization privileges immediately. This means getting back or invalidating all physical access cards, keys, etc., and immediately deleting or invalidating their passwords and other authorizations from every system they ever had access to. It is especially important to remove their ability for remote access (through modem, virtual private network, etc.). If they had access to any group or shared accounts, those passwords should be changed immediately.

Applying the principle of least privilege in practice is difficult, if not impossible, without the right access control technology. The different types of access control technologies are covered in Chapter 7. Chapter 7 discusses role-based access control, an important technology to enable adoption of the principle of least privilege, as well as to simplify and better manage identification, authentication, and authorization.

## 6.4   Network Separation

Network separation is a perimeter or boundary defense, which we discussed in Section 6-2. Let's look back at Figure 6-4, Cyber Defense in Layers, and look at the connection between our office LAN, in Layer One, and the Basic Process Control System (BPCS).

The principle of defense in layers implies that a direct office LAN-to-industrial network connection is not a good idea. Anyone having access to the office LAN, whether access was obtained legitimately or illegally, now has complete access to the industrial network and its components, including HMIs, control servers, etc.

So what should our risk team do about a direct business-to-control system connection, if it exists?

Applying the basic risk management choices detailed in Chapter 5-1, the risk team may elect to:

1.  accept the risk, and do nothing, leaving a direct connection to the industrial network;

2.  partially close off this access with a firewall, filtering router, or other restriction; or

3. cut the connection between the business and industrial networks completely.

Most companies in the chemical processing, utility, and discrete manufacturing industries say they need some connectivity between the business network and industrial network to survive. There is just too much business advantage from having some form of connectivity and information flow.

In the writer's experience, most companies started out with an unfettered business-to-industrial network connection. While some continue to elect Option 1, accept the risk, most are going to Option 2, putting in an internal firewall or other network restriction such as a filtering router.

Chapter 10 presents an account of the way a large company has handled internal business-to-control system connections.

Few companies will elect Option 3, to cut the connection. However, some companies that never connected the industrial and business networks to begin with may continue to observe that policy.

# References

1. U.S. Chemical Safety and Hazard Investigation Board *Investigation Report – Chemical Manufacturing Incident*, Report No. 1998-06-I-NJ. (April 8, 1998). Retrieved 11/11/2004 from: http://www.csb.gov/Completed_Investigations/docs/Final%20Morton%20Report.pdf

2. American Institute of Chemical Engineers (AIChE), Center for Chemical Process Safety. *Guidelines for Safe Automation of Chemical Processes*. AIChE, 1993.

3. American Institute of Chemical Engineers (AIChE), Center for Chemical Process Safety. *Guidelines for Safe Automation of Chemical Processes,* Figure 2-2. AIChE, 1993.

4. ANSI/ISA-99.00.01-2007, *Security for Industrial Automation and Control Systems, Part 1.* Research Triangle Park, ISA, 2007.

# 7.0
# Cyberdefense Part II – Technology

## 7.1    Guidance from ISA-99 TR1

The ANSI/ISA-TR99.00.01-2007 – Security Technologies for Industrial Automation and Control Systems standard has a wealth of information on IT security technology and how it may be applied to securing industrial networks. Each technology is summarized according to the following headings:

- Security Vulnerabilities Addressed by this Technology, Tools and/or Countermeasures
- Typical Deployment
- Known Issues and Weaknesses
- Assessment for Use in the IACS Environment Systems
- Future Directions
- Recommendations and Guidance
- Information Sources and Reference Material

The sections in this chapter cover some of the technologies described in the ISA-99 series of standards. Our coverage of these technologies is intended to be a general introduction to the various technologies and how they are used, rather than a detailed technical explanation.

## 7.2    Firewalls and Boundary Protection

A firewall acts as a "gatekeeper" or "traffic cop" to filter and block traffic from one network going to another. Let's look at two cases, illustrated in Figure 7-1:

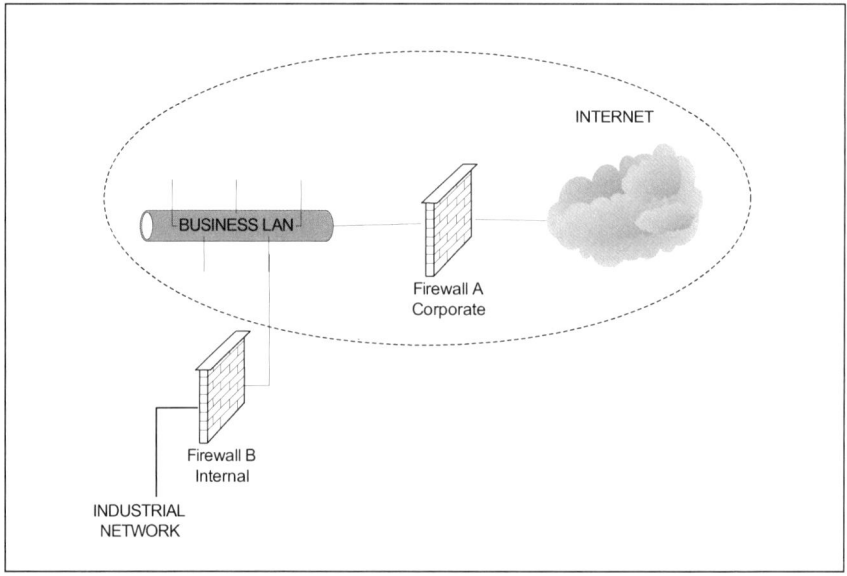

**Figure 7-1. Firewall Illustration**

- Firewall "A" protects the corporation business LAN from the outside Internet.
- Firewall "B" is internal and separates the business LAN from the industrial network.

Each firewall has a set of firewall "policies" (not to be confused with the higher-level security policies described in Chapter 8) that determines which hosts or networks on one side may talk to hosts or networks on the other side.

It all boils down to a yes/no decision for each, whether to permit or deny each attempted connection.

As an example, let's look at classes of users inside and outside the business network, as shown in Figure 7-2, and what connections they might want to establish.

| SOURCE | DESTINATION | SERVICE | PERMIT OR DENY |
|---|---|---|---|
| Business LAN User | Outside Internet | Web (http) | Permit |
| Business LAN User | Outside Internet | RealAudio | Deny |
| Outside Internet Host | Business LAN Host | Web (http) | Deny |

**Figure 7-2. Sample Firewall Setup**

If a business LAN user wants to connect to an outside web server (the firewall "listens" for attempts at connection via the web protocol known as HTTP), this is "permitted" (unless management is clamping down on too much outside web surfing!)

However, if a business LAN user wants to connect to an outside streaming "RealAudio" server, perhaps this connection will be "denied" by Corporate IT cybersecurity.

Let's take a look at attempted traffic going the opposite direction. If a machine on the outside, host "hacker.com," wants to connect from the outside Internet to an inside business LAN workstation or server, this should be blocked or "denied." Most corporations host a web server in an intermediate zone called a DMZ (Demilitarized Zone) for legitimate incoming traffic such as to get sales bulletins and the like.

ISA-99 TR1 goes on to describe three different types of firewalls:

- Packet Filter
- Application Proxy
- Stateful Inspection

Modern firewalls may be hardware-based (e.g., a firewall appliance with embedded software) or software-based, running as application software on a Windows or Unix operating system. If software-based firewalls are used, the underlying operating system must be hardened, as described in Chapter 8, to be effective.

An example of a modern chemical corporation using internal firewalls is given in Chapter 9.

## Alternate Internal Boundary Protection

Nearly all corporations will have a corporate firewall (Firewall A as shown in Figure 7-1). However, some may elect not to go with a full-fledged internal firewall (Firewall B in the figure) to separate critical internal systems from their business LANs and intranets. A degree of protection can be provided by using a router with filtering capabilities. For instance, using a router's Access Control Lists (ACLs), a network administrator can select which hosts and networks on one side of the router can connect with specific hosts and networks on the other side of the router, as described earlier in this section in the discussion of firewall policies.

# 7.3   Intrusion Detection

Intrusion detectors monitor computer networks or computer hosts, looking for possible intrusions. There are two general types of intrusion detectors:

- Network-based (NIDS – Network Intrusion Detection System)
- Host-based (HIDS – Host Intrusion Detection System)

A network-based intrusion detector may be attached to the network it monitors by a "network sniffer" arrangement, or it may be embedded into the operating code of a router, firewall, or standalone appliance.

It may look for either or both of the following warning signs:

- Known attack signatures, recognized from an up-to-date database of known attacks such as worms.
- Network traffic anomalies, changes in traffic patterns that are statistically suspicious. For instance, heavy incoming traffic on a little-used port or IP address might indicate an attack.

A host-based intrusion detector is mounted on a particular host computer, such as a workstation or server. It may perform a periodic scan of all crucial files on the host to look for signs of unauthorized alteration, which might indicate a compromise of the host system by an intruder. This action is called a "file integrity check." It may also monitor network traffic in and out of a particular host, or look for suspicious usage patterns, which might indicate an intruder is at work.

Figure 7-3 shows how a typical NIDS and HIDS might be deployed in the corporate network example displayed in Figure 7-1.

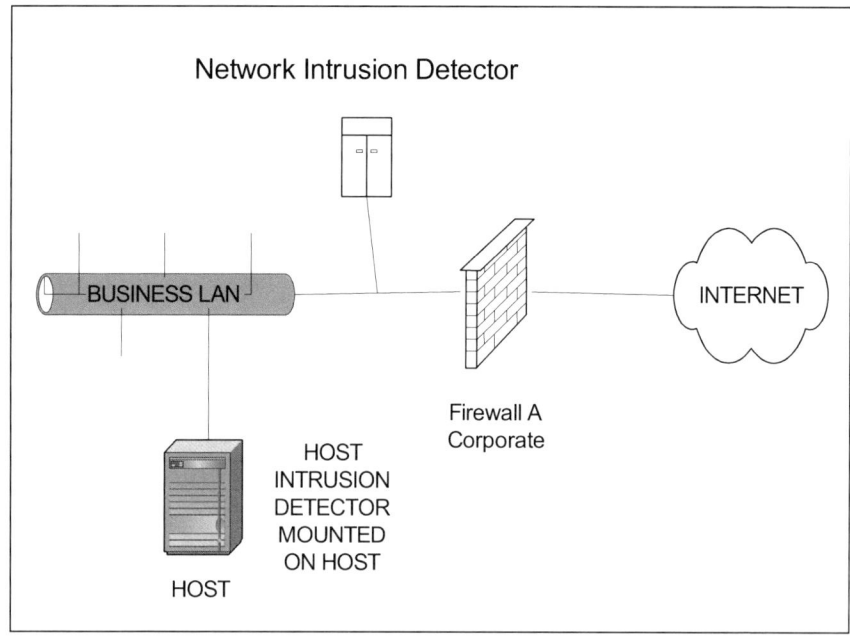

Figure 7-3. Intrusion Detection

Figure 7-3 shows the NIDS deployed to listen to or "sniff" the network traffic just inside the corporate firewall. It looks for signatures or patterns of intrusion from the outside Internet past the corporate firewall.

On the other hand, the HIDS monitors one host; in this case, the host on the business LAN.

The action taken by a NIDS or HIDS upon sensing a potential break-in can vary, anywhere from sending an email to paging a system administrator.

An emerging variation on intrusion detection is called intrusion prevention. This detector automatically takes a prearranged action upon any sign of intrusion. For instance, if the NIDS in Figure 7-3 were to detect an anomaly and cause the firewall to block some or all traffic into the business network from the Internet, it would be actively doing intrusion prevention rather than the more passive notification that comes with intrusion detection.

One concern with deploying NIDS and HIDS is the tendency for false alarms, or false positives, which take time and effort to track down. Just as you don't want a burglar alarm to go off because it thinks the family pet is a burglar, minimizing false alarms is necessary when deploying this technology.

# 7.4   Virus Control

Since the advent of the PC, there has been a constant struggle between virus writers and people who make software to detect and control viruses. Over the years, new and more clever viruses have evolved, and antivirus researchers are evolving more strategies to spot and clean them.

The virus prevention and detection cycle is a "chase your tail" game. More than 50,000 viruses are known to exist. A large number of them are "zoo" viruses, which exist in controlled laboratory collections only. As we are only too aware, however, a significant number of "in the wild" viruses have been released into cyberspace and have done damage.

Figure 7-4 shows the dilemma antivirus researchers face.

Figure 7-4 illustrates a situation in which a virus writer creates a totally new virus, or a new variation on an old virus, and releases it "in the wild." Some computers get infected, and their owners send a sample of the new viral infection to an antivirus vendor's research team.

Within a few hours, the antivirus team has "disassembled" the inner workings of the virus and captured that virus's distinct signature, or code pattern, as a short sequence of bits. The antivirus vendor then dis-

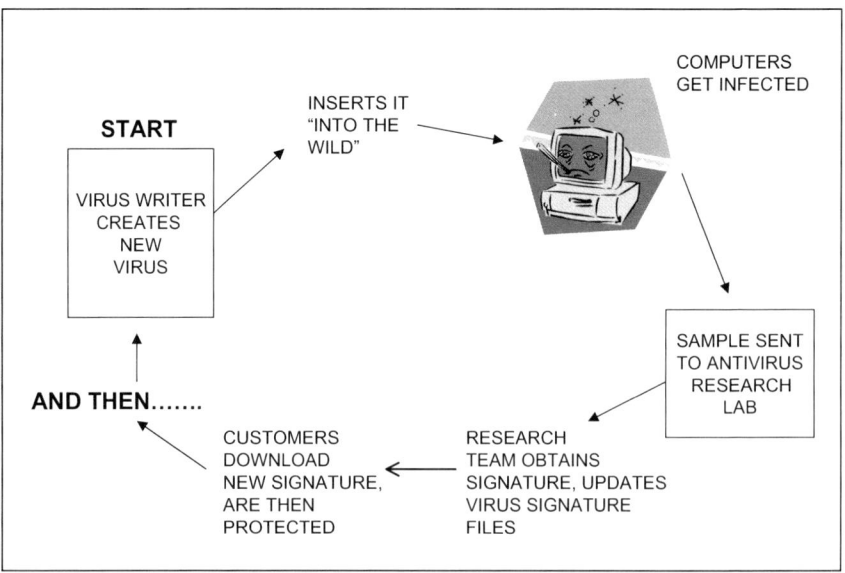

**Figure 7-4. The Antivirus Cycle**

tributes that virus signature to its customers as an update of their virus signatures file.

The problem is that the virus signature they developed is valid only for that particular virus. Virus writers can "tweak" a virus to alter its code pattern and make a new version that will go undetected. Virus writers may go as far as buying several brands of virus detection software in order to download the latest signature file updates and check to see if their "tweaked" virus is detectable!

Thus, there is a constant running battle between virus writers and the antivirus research community.

Several antivirus products try to detect new viruses for which no signature is yet available. This antivirus software watches for unusual program behavior or combinations of behaviors in an effort to identify viruses up front, before infection.

Antivirus programs typically contain three parts:

1. *The Graphical User Interface (GUI).*

2. *The Engine.* This contains the scanning software, which compares files on the host computer with the latest virus signatures from the signature file.

3. *The Signature File.* Downloaded at regular intervals, say each day, it contains signatures of the latest viruses and Trojans.

Viruses may attack various locations in operating programs and memory. Figure 7-5 shows just a few of the major viruses that have attacked in history, along with the type of attack.

| VIRUS | DATE | TYPE |
|---|---|---|
| Lehigh | 1987 | MS-DOS File Infector |
| Concept | 1996 | Macrovirus |
| I Love You | 2000 | E-Mail |

**Figure 7-5. Some Past Virus Attacks**

## Some Past Virus Attacks

Virus detection and/or elimination may be deployed at three levels, or tiers, within the industrial network:

• At the perimeter of the industrial network. Virus protection may be built into or added onto firewall products.

- At the control server level. Server editions of antivirus products may be used here.

- At the individual workstation or PC level. For instance, the workstation running the HMI console may have antivirus software to protect against employees bringing in diskettes, flash drives, or CDs with viruses.

At present, there is still some residual discussion about whether using antivirus software at the control server or workstation level will interfere with proper operation. Many control vendors approve using only specific brands of antivirus software that have been tested for non-interference with application software. In addition, the vendors may specify that only certain features of the antivirus software may be used, and it must be configured a certain way.

In 2006 a report titled, "Using Host-Based Antivirus Software on Industrial Control Systems" was issued, describing the results of a two-year DOE National SCADA Test Bed study written on the subject of using host-based antivirus software on control systems, written by the author, Steve Hurd, and Joe Falco from NIST[1].

If a virus is detected in real time, the next question is: What is the plan to isolate the network section, clean the virus, and then get back in operation? This is part of an incident response plan that must be set up.

# 7.5   Encryption Technologies

Encryption technologies are the practical application of the field of cryptography, which means "secret writing." Cryptography has been used in many forms since ancient times to conceal information lest it fall into the wrong hands. A message, once encrypted, appears as gibberish and is of no use to an adversary unless the adversary knows how to reverse or decrypt the encrypted message.

To understand the basics of encryption, some terms need to be introduced:

- *Plaintext.* The "plain English" version of a text or numerical message to be concealed.

- *Ciphertext.* The plaintext transformed by an encryption algorithm, using an encryption key, into a message that is unreadable without being decrypted.

- *Encryption Algorithm.* The mathematical formula or procedure or other formula that will convert the plaintext to ciphertext.

- *Encryption Key.* A unique combination of numbers and/or digits that is used by the encryption algorithm to convert plaintext to ciphertext.

Let's give a simple example of the use of an encryption algorithm with key, attributed to Julius Caesar and his method of "secret writing." The Caesar cipher uses a very simple secret key algorithm, called a substitution cipher. We substitute new letters for each letter of original text to make the original text illegible.

Suppose we're communicating with the battlefield, and the message we want to send is:

ATTACK AT DAWN

Our encryption algorithm works as follows: First we write out the letters of the alphabet. Then we write out a second alphabet beneath the first alphabet, except we shift it one letter over:

ABCDEFGHIJKLMNOPQRSTUVWXYZ
ABCDEFGHIJKLMNOPQRSTUVWXY

Starting from the bottom alphabet, wherever we have an A in our original message, we look directly above it and substitute a B, in the top (shifted) alphabet. So our original message ATTACK AT DAWN becomes the unreadable

BUUBDL BU EBXO

(In practice, we can eliminate the spaces between words as well.)

The key to our simple alphabet substitution algorithm is the number 1. We shifted the alphabet over by one letter to form ciphertext. We could just as easily have shifted the alphabet by 2, so that A would now become C, B would become D, etc.

Caesar's general in the field, receiving the cryptic message BUUBDL BU EBXO only needs to know the algorithm and the key to get back the plaintext ATTACK AT DAWN. Using the two alphabets above, the general goes from top alphabet to bottom, reversing the way the encryption was performed.

The "key space" is the number of unique values the key can take. What are possible values of the key? Well, we can shift the alphabet by up to the number of letters in the alphabet, 25. (If we shift 26, we circle around the alphabet and come back to where we started.) So we have 25 unique keys that can be used with this simple substitution algorithm.

If the enemy finds out the algorithm being used is the Caesar cipher, he can try a brute force attack against the algorithm, using one message in the ciphertext he has managed to intercept: BUUBDL BU EXBO.

By trying each unique combination in the key space, 1-25, the enemy can discover the key used. In our example, if he just tries the number one, the plaintext becomes evident.

As has been mentioned, the Caesar algorithm is called a secret key algorithm. Only the sender and recipient of the message may know the secret key. If an adversary finds out, all is lost.

Writing secure cryptographic algorithms is very difficult. The algorithm must be resistant to an attack by analysis, called cryptanalysis. And the key space must be large enough that it would take too long to find the key through trial and error (a brute force attack).

In our example, if dawn and the attack come before the adversary can find the right key by trial and error or any other method, then the algorithm will have served its purpose.

Modern-day secret key algorithms use mathematical calculations with key sizes described in terms of bits. The Data Encryption Standard (DES) algorithm, which is at the end of its useful life, uses 56 bits. A brute force attack on DES is very time consuming but achievable with today's computing power. It is being superseded by the Advanced Encryption Standard (AES), which uses up to a 256-bit key.

Just like the cat-and-mouse competition between virus writers and antivirus researchers, there is a running competition between cryptographers, who develop new encryption algorithms, and practitioners of cryptanalysis, who try to break them by many different means. At stake are billions of dollars—for instance, in interbank money transfers that might be compromised if someone on the wrong side discovers the key or how to crack the algorithm.

## Public Key vs. Secret Key Algorithms

Secret key algorithms, running the gamut from the Caesar cipher to DES and AES algorithms, are designed to preserve confidentiality. (Remember the AIC triad outlined in Chapter 6?) The confidentiality of the data (plaintext) is preserved only as long as the adversary does not have access to, or the ability to figure out, the secret key by a brute force attack or any other method.

Another form of cryptography, public key cryptography, was invented in 1978 by three individuals, for whom it is called RSA: Rivest, Shamir, and Adelman. It may be used for both authentication and confidentiality.

In public key cryptography each user has two keys, or a "key pair." A key pair is made up of a public key, which may be given out in "public places," and a private key, which must be kept secret by the user. The two keys are mathematically related. Figure 7-6 shows how public key cryptography may be used to ensure confidentiality.

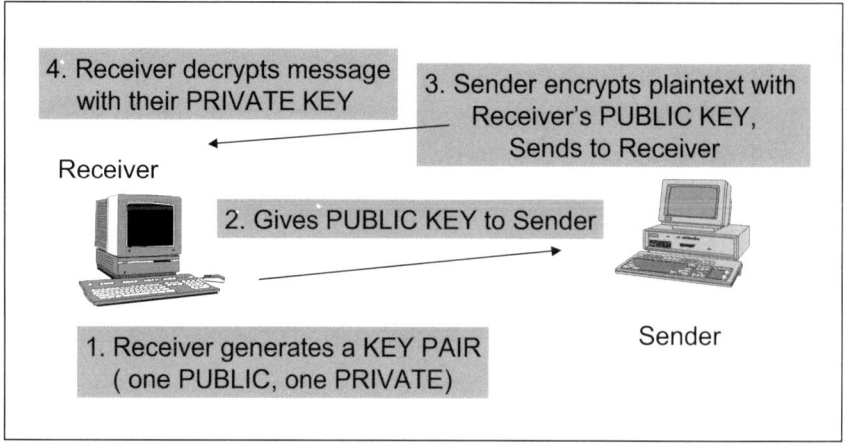

**Figure 7-6. Using Public Key for Confidentiality**

Referring to Figure 7-6, the receiver generates a key pair and keeps the private key secret, but sends the public key to the sender, who wants to send the receiver a confidential message.

The sender encrypts a plaintext message with the receiver's public key, then sends the encrypted message back to the receiver. The receiver, using the private key, is the only one who can decrypt the message.

This illustration shows we can use a public key algorithm to do the same thing as a secret key algorithm. In practice, though, using a public key algorithm takes much more processing time. It would not be practical to use public key to encrypt and send large amounts of data. In practice the public key is used in combination with a secret key for this purpose.

The real advantage of public key encryption is that it may be used for authentication.

Figure 7-7 shows how we may have our users authenticate each other.

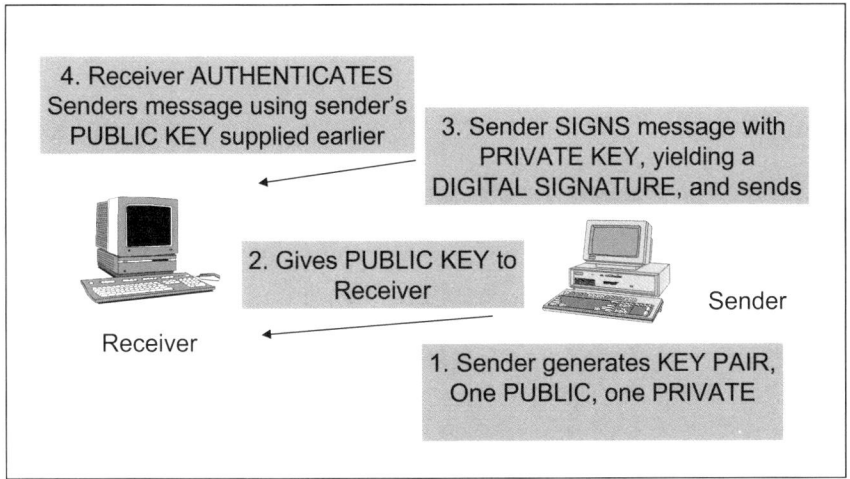

**Figure 7-7. Using Public Key for Authentication**

Referring to Figure 7-7, suppose the receiver wants to be sure the message really came from the sender, not an imposter. If the sender and receiver had each generated their own key pairs and then swapped public keys, this would be achievable. The receiver would have the sender's public key to begin with. The receiver would ask the sender to "sign" the message with his or her private key, creating a digital signature. Upon receiving the message, the receiver would check the sender's digital signature against their copy of the sender's public key to see if they matched. If they did, the message indeed came from the real sender, not an imposter.

As we can see from the above example, if two users generate key pairs, they may be used for both authentication (digital signature) and confidentiality (encryption).

In our previous example, the sender and receiver have met in person, know each other, and, therefore, have a "trust relationship." But what if the sender and receiver have never met and established that trust relationship? How does the receiver know the public key received originally from the sender really belongs to the sender and not to an imposter?

The answer is to provide a public key infrastructure, or a way of certifying or guaranteeing the public keys are genuine and really belong to the authentic senders. This is usually done by an outside agency such as a bank or other certifying agency. The outside agency certifies in some way to the receiver that the sender is authentic (by requiring proof of identity, for instance) and the public key is genuine.

**Message Integrity Checking**

We need another type of cryptographic algorithm to complete our crypto toolkit–an algorithm that can let us know if a message has been altered in any way. A cryptographic checksum does this for us. Using an algorithm, it sums up the unique pattern of ones and zeroes comprising the binary representation of a message, generating a short checksum.

In telecommunications, a cyclic redundancy check (CRC) is used for this purpose–after every frame of data a cyclic redundancy check is computed and tacked onto the end of the message. Computing a cryptographic checksum ensures that the message/checksum correspondence cannot be tampered with.

Adding a cryptographic checksum to our toolkit gives us methods to ensure confidentiality, authentication, and message integrity.

### Application of Cryptography to Industrial Network Security

Applications using cryptography are entering the field of industrial network security at a slow pace for the following reasons:

1. Encryption is a complex subject and requires an understanding of the mathematical basis of the algorithms used.

2. Adding encryption to industrial network data transmissions adds processing time to what may be fully utilized microprocessors and also requires additional communications bandwidth. When talking about response time in milliseconds or for deterministic control applications, the latency or "jitter" introduced could delay crucial control events.

3. Key management. Generating, storing, and distributing keys can be a difficult process. If using public key infrastructure (PKI), a suitable structure must be set up.

# 7.6   Virtual Private Networks (VPNs)

Virtual private networks fulfill an important role in the networked world and the Internet.

Using the open Internet, they are designed to give protection to data communication equal to or greater than sending data via a dedicated phone line. A VPN works by setting up a secure tunnel over the Internet using an encrypted connection, and offers these three capabilities:

1. Identification, Authentication, and Authorization (see 7.7)

2. Integrity of information transfer

3. Confidentiality

Figures 7-8 and 7-9 show two ways a VPN might be set up.

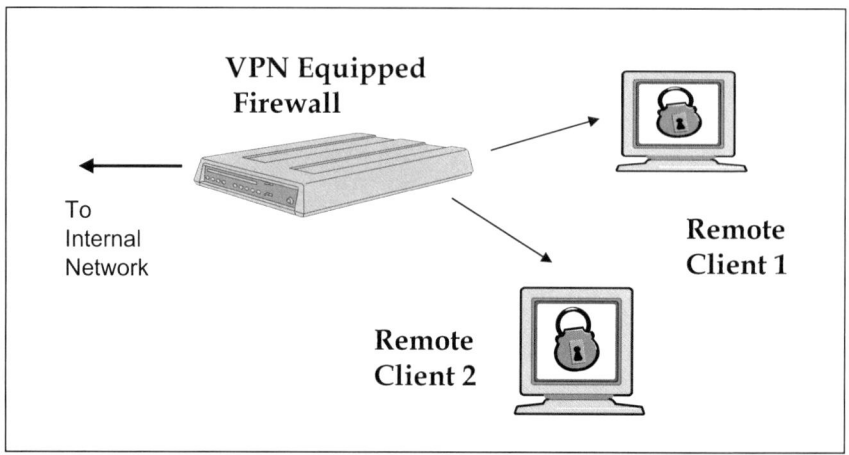

**Figure 7-8. VPN Configuration 1**

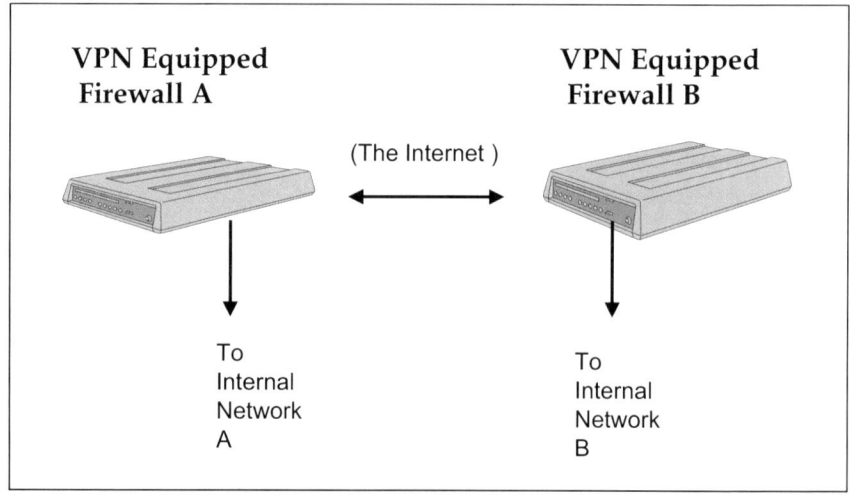

**Figure 7-9. VPN Configuration 2**

Figure 7-8 shows a VPN configuration for giving secure remote access across the Internet. Here, remote hosts (say two different employees working at home) may access a corporate private network securely by setting up VPNs to their laptop computers. They would log into their local Internet Service Providers (ISPs), go to the web address set up for their corporation's VPN equipped firewall, authenticate themselves, and be granted access.

In the configuration shown in Figure 7-9, the VPN connection allows private network A, shielded from the Internet by Firewall A, to connect securely with private network B, which is similarly shielded from the open Internet by Firewall B.

# 7.7 Authentication and Authorization Technologies

In Section 4.3 we dealt with the issues of Identification, Authentication, and Authorization. We introduced these concepts as follows:

- Identification = Who are you?

- Authentication = Prove it.

- Authorization = Now that we've established your identity, what set of access privileges do you have?

We also introduced the three factors of authentication as the following:

- Something you know

- Something you have

- Something you are

We can use any factor of authentication alone or in combination with other authentication factors to have a stronger authentication.

In cyberspace, using something you know translates into using a password or pass phrase. A password is relatively short, say eight alphanumeric characters, and a pass phrase is longer. This is the most time-honored and widely used method of cyber authentication. This method assumes the system user will enter a secret and cryptic combination of letters and/or numbers, and then will remember them the next time he or she wants to log onto the system.

Anyone not knowing this cryptic combination of letters and numbers would have to get the password from the user by trickery somehow or resort to brute force guessing, a trial-and-error method of testing all possible combinations of numbers and letters that might make up a password or pass phrase.

To be effective, passwords or pass phrases must:

- Have enough characters so the task of a brute force trial-and-error attack would be prohibitively time-consuming;
- Not be easily guessable by another party;
- Be retained in the user's memory only, not written down on slips of paper, sticky notes, etc.; and
- Be changed at reasonable and regular intervals, say once or twice per month.

Authentication with "something you have" equates to authentication with a key or hardware token. One of the most direct ways to provide authentication is by resorting to a physical security device, such as a lock, with a key carried by the user.

The user plugs in a hardware token to gain access, perhaps one in the form of an Radio Frequency Identification Device (RFID) or a USB dongle. An embedded-chip card or a system using a magnetic stripe may be used also.

Authentication with "something you are" brings up the rapidly developing area of biometrics—the technology of verifying identity with a unique physical attribute that is not easily duplicated. Biometric identification can include the following:

- Hand Geometry
- Fingerprint
- Voiceprint
- Face Recognition
- Signature Recognition
- Iris Recognition

The field of biometrics has come a long way in the last few years. Some of the above methods, such as hand geometry, have been used in industry for 20–30 years; others, such as face recognition, are much newer.

Biometrics may be abused as well as used properly.

When system developers have tried to use biometrics for identification and authentication together, rather than for authentication alone, they have generally not been successful. Reference (2) is a news story of an attempt to use face recognition to catch criminals by the Tampa, Florida, police department that failed to produce results.

## Increasing the Factors of Authentication

Greater confidence in the authentication process may be had by using two or more factors of authentication, either multiple instances of the same factor or different factors. For example, in a popular two-factor authentication process referred to in Section 4.3, a token flashing a one-time password that changes each minute can be used as a centralized log-in screen, where the user must input a pass phrase consisting of a unique four-character PIN that doesn't change (something you know) with the one-time password (also something you know) displayed on the encryption token to log on and get access to the computing services.

## Authorization

Finally, let's talk about authorization. As introduced in Section 4.3, once a user (or device) is identified and authenticated, we need some way of allocating certain access privileges to the person or device. What are they permitted to do? Which files may they change, delete, or create?

Historically, several conceptual models of authorization have been used by government and the military, and by industry.

- *Mandatory Access Control.* This has been used in military and government circles. Here information files are classified "Secret," "Top Secret," etc., and only persons with the matching secret or top secret security clearance may have access to these files. Control is centralized, and based on a rigid set of access control rules.

- *Discretionary Access Control.* This has been used commonly in industry and commercial computer systems. Here, whoever "owns" the information is empowered to set limits on who may access the information and what privileges they have to modify it.

- *Role-Based Access Control.* This type of access control shows great promise for industrial networking situations. Here, the users are grouped into roles, depending on what their job function is. For instance, in a bank, the roles might be teller, head teller, branch manager etc., with a number of individuals belonging to a role group. Once employees are identified and authenticated within the system, their roles determine their authorization privileges, not their individual identities. One can see the efficiency advantage if, for instance, a centralized role-based access control system were used in a large industrial control room. Operators, shift supervisors, engineers, and technicians would each be in a role group that would have certain fixed privileges. If one employee leaves and another arrives, each only needs to add or delete their individual identities to the roles list on the centralized server, not add or delete them from access control lists on pieces of individual systems in the control list.

It should be emphasized that identification, authentication, and authorization don't pertain exclusively to people. A secure intelligent device, such as a control sensor or actuator or a PLC on a network, may need to identify itself to the rest of the control network as the "real thing" and not an "imposter device." And a whole subnetwork (for instance, a remote industrial network segment) may need to identify itself to another network. Identification, authentication, and authorization are for machines, devices, and industrial network segments as well as for people.

# References

1. Falco, J., Hurd, S., and Teumim, D. "Using Host-Based Antivirus Software on Industrial Control Systems." *NIST Special Publication 1058* (2006).

2. Bowman, L. M. *"Tampa Drops Face-Recognition System."* Cnet.com article. August 21, 2003. Retrieved 11/11/2004 from: http://news.com.com/Tampa+drops +facerecognition+ system/2100-1029_3-5066795.html

# 8.0
# Cyberdefense Part III – People, Policies, and Security Assurance

## 8.1 Management Actions and Responsibility

In Chapter 2, we saw that to be effective, industrial network security has to be driven by top management and work its way down the corporation. The alternative, a "grass-roots" effort by automation and control engineering, may be commendable but will probably not get the attention and resources it needs to succeed in a measurable way.

Several key factors are necessary to develop a meaningful industrial network security organization and program. Two of these factors are:

- Leadership commitment. Industrial network security needs a genuine place in the organization, a place that fits in with corporate goals for risk management and for corporate and IT security. This means top management must be committed, and this often means a convincing business case must first be made (see Chapter 2).
- An industrial network security committee, task force, or similar entity. This entity may be called a Program Team.

Resources for the Program Team must include:

- Personnel
- Budget
- Training

- Organizational empowerment and authority
- A charter, usually some high-level security policies that detail the mission, structure, goals, and responsibilities of the Program Team
- A first project—as modest or as ambitious as Program Team resources will allow
- A plan for the first project

## 8.2  Writing Effective Security Documentation

Security documentation creates a vehicle for informing your company about recommended and/or required practices for cybersecurity that can be read and understood by readers at all levels of technical sophistication. Most readers want to spend as little time as possible wading through information that does not apply to them to get to what they really need.

Let's talk about IT cybersecurity before we consider industrial networks. There are many different approaches to writing security documents in the IT world, and the resulting documentation may be labeled differently and be composed of different sets of information from company to company.

The writer's point of view, after spending many hours in fruitless discussions with peers over which piece of paper should be called by what name, is that the issue is not so much what name we give to our documents but whether the documents, taken together, convey the required information in an efficient fashion. Also, does the final set of security documents "hang together" and produce a coherent framework for the various readers?

With this introduction in mind, let's look at the business side of the company we described in Chapter 2. A set of IT cybersecurity documents for the business side of our widget factory would address these issues, among many others:

- *Web.* Downloading of pornography or other illegal content by employees.
- *Email.* Viruses and spam coming in with email.
- *Remote access.* Allowing authorized users to connect via modem or VPN and keeping hackers out.
- *Unlicensed software.* Keeping employees from using unpaid-for software.

What sort of security documentation system is best to convey all the required security information? The writer presents the following IT cybersecurity framework as one system that "hangs together." By no means is it the only way to also structure a set of industrial network security documents, but it is a common and proven way.

This system uses four types of security documents:

- Security Policies
- Security Standards
- Security Guidelines
- Security Procedures

Classification of security documents into the categories above depends on the message, the intended audience, the document's technical sophistication, and whether the message and instructions are recommended or mandatory.

Let's start at the top of the list. Security policy usually comes from high in the management chain and is a short statement of the corporation's position on security issues. For instance, it may come from as high a level as the CEO of the company, saying something such as, "This corporation believes that IT cybersecurity is crucial to the success of the company for the following reasons: (list reasons). Therefore, we have assigned the (name of group), under the leadership of (name or title of person in charge), to be responsible for this area and to report to me at regular intervals."

Among IT cybersecurity professionals, the term "security policy" may also be used at much lower levels. For instance, the security policy for a firewall may simply be a list of rules for setting up a firewall. Among IT professionals this may be an allowable use for "security policy," but we must clearly differentiate this document from the CEO's proclamation!

We will show how to do this in an upcoming figure. Let's now define the three other security documents listed above:

- *Security Standard.* A document that is mandatory and prescriptive, describing how to deal with cybersecurity issues. For example, "A firewall must be used at every connection from the business LAN to the Internet." It may also include provisions such as the level of approval necessary for elements of the system not to be subject to a certain part of the requirement.
- *Security Guidelines.* A document that describes recommended but not mandatory ways to solve security problems or sets forth options for solving problems.
- *Security Procedures.* Detailed technical documents for accomplishing security tasks and meant for the employees doing the work. A security procedure may be a mandatory or recommended way to perform a security task.

Next, let's create a framework on which hang the four types of security documents while allowing for different levels of security policy. Figure 8-1 gives such a security document framework.

As shown in Figure 8-1, security policies cascade from the highest level (CEO level) to mid-level (CIO or IT cybersecurity) to low level (for

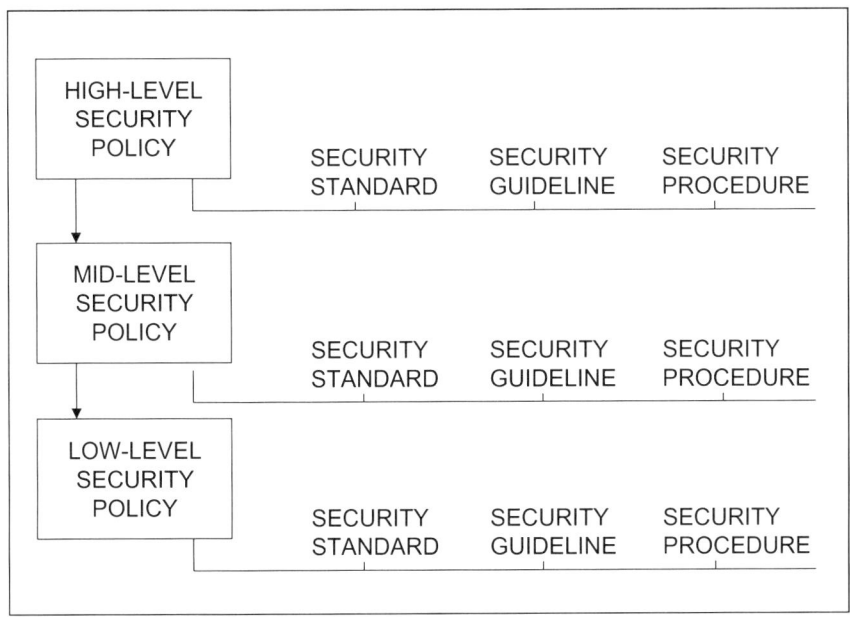

**Figure 8-1. A Cybersecurity Document Framework**

instance, the industrial network security level). The aforementioned Program Team that decides and implements security within the industrial network boundary might be an excellent choice to write the low-level security policies.

Consider a specific example from our list of typical IT cybersecurity issues–Internet and email use by employees. At the top (CEO) level, there might be policies on "business only" use of Internet and email by employees. At mid-level (CIO), there might be further policy qualification of what constitutes business-only use of these resources, with standards, guidelines, and procedures to enable and enforce this policy.

Finally, the low-level policy describes how Internet and email access will be addressed inside the industrial network boundary.

A major cybersecurity question may be whether to allow company email and Internet connectivity to any computer connected to the process control network, for fear of spreading viruses or Trojan horses to critical process networks.

Some alternatives might be to:

1. allow company email and Internet connectivity to any operator or engineering workstation, as desired;
2. allow company email and Internet connectivity only to certain controlled and monitored workstations; or
3. not allow any company email or Internet connectivity to any computer on the process control network. (This is the most restrictive security policy, and the approach favored by the writer.)

However, an alternate means of providing email and Internet access within the control room is to extend the business LAN into the control room as a parallel, "air-gapped" network, and have dedicated business workstations for operators. This way, business network connectivity is provided without direct process control network access.

But let's say alternative 2 is chosen. The security documents might be framed around the mechanism and infrastructure to provide this solution.

The Security Policy would simply state that only certain designated and controlled workstations on the process control network could be used for Internet and email.

A Security Standard might specify the type and number of workstation allowed, who will set these up, the configuration, method of monitoring, auditing, etc.

A Security Procedure might be the instructions to the IT/Control Engineering staff on exactly how to set up these workstations.

A key feature of the security document framework is that one group of readers is not burdened with unnecessary detail meant for another group of readers. The policy document has no need for the technical details of how to set up the workstation. This security document framework is modular, concise, and provides for different documents for different classes of readers.

## 8.3   Awareness and Training

One area of security that is frequently overlooked is industrial network security awareness and training for all the users of a system or group of systems.

Security awareness is accomplished when industrial network users understand the need for security, the threats and vulnerabilities in a general way, the security countermeasures and why they are designed the way they are, and how the lack of secure operation of these systems will affect their jobs and the company's bottom line.

It is important to repeat awareness sessions to regularly remind employees, contractors, and other users of the system of these matters and to keep them up to date on changes.

Some formats for awareness sessions with employees might be:

- Live security talks or presentations
- Printed materials, such as brochures, posters, etc.

The security awareness program is for everybody–all who will use or come in contact with the systems. On the other hand, security training is specific. Security topics may be presented in self-taught sessions or in more formal classroom sessions. For instance, training new engineers on the method for secure remote access over a VPN might be a suitable topic for a "hands-on" training session.

## 8.4   Industrial Network Security Assurance Program: Security Checklists

Security checklists are lists of routine activities that must be completed to accomplish a certain security goal, such as securing a host or network. They are used extensively for day-to-day activities in IT cybersecurity and may also be used for industrial network security tasks. Let's look at some functions security checklists provide in IT cybersecurity.

One way COTS software can be vulnerable to cyberattack is by having open ports and services on the host computer that aren't being used, thereby opening avenues of attack. This is much like leaving many doors in a big building unlocked even though no one uses these doors.

COTS operating systems, when installed "out of the box," frequently leave services (from web servers to exotic, little-used services) and ports open by default. It is the opposite of the basic security principle–the Principle of Least Privilege–described previously. If ports and services are not closed in a systematic procedure, these open doors make cyber-attack easier.

Another way COTS software may invite cyber attack is by leaving unpatched vulnerabilities. As discussed previously, many vulnerabilities in COTS software for business and industrial network applications are coded into the software during the development process and then not caught in a code inspection or quality assurance effort before release. We saw in Chapter 4 that a simple buffer overflow condition is responsible for many security vulnerabilities.

Unfortunately, these vulnerabilities are then found one at a time by security researchers or by the hacking community. If a vulnerability is caught by a security researcher, perhaps after a user complaint, the researcher should work with the vendor to ensure that a patch is developed and available at the same time as the vulnerability is made public.

This gives conscientious system administrators time to download the patch from the vendor's web site and fix their systems, hopefully before a new virus or worm targeting that vulnerability can be invented by a hacker.

Vendors and non-profit security organizations have security checklists and even automated system configuration tools to identify and close the unneeded ports and services described above, as well as to check on security patch level and installation, in a step-by-step fashion.

This process of patching vulnerabilities and turning off unneeded ports and services for your computers and network equipment is known as "host and network hardening."

An example of a coordinated host and network security hardening project is a program begun in 2003 by the National Institute of Standards and Technology (NIST). NIST began to gather and put into a database many different security checklists and automated configuration toolsets furnished by such companies and organizations as Microsoft, the National Security Agency (NSA), and others.[1]

The concept of host and network hardening and security checklists may also be applied to industrial network security. Some applications might include:

- checking an industrial network security configuration before putting it into production mode or
- hardening a Windows or Unix host before connecting it to an industrial network.

Before using an IT security checklist for an industrial network, one additional step is necessary: letting the industrial network vendor review and test the checklist activities, including closing ports and services and applying patches, to ensure that checklist activities are compatible with the application software as installed. Figure 8-2 gives a simple flowchart that includes this extra step.

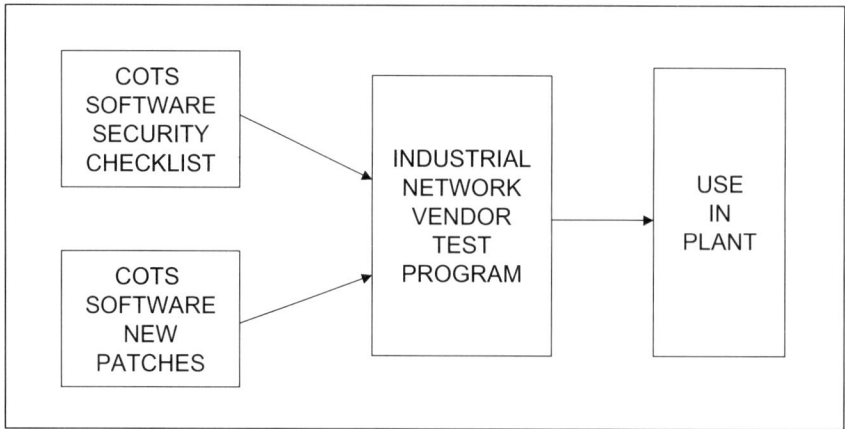

**Figure 8-2. Industrial Network Hardening Flowchart**

Once "blessed" by the industrial network vendor as in Figure 8-2, security checklists may be very easily incorporated into the security document framework outlined previously, at the level of standards, guidelines, or procedures. They will save time, improve uniformity and consistency of security efforts, and help ensure that organizational knowledge of industrial network security is not lost if key people leave the company.

# 8.5   Security Assurance: Audits

Security audits are also frequently used in IT cybersecurity as a means of:

- checking that changes to a network's setup and configuration are satisfactory and agree with established security procedures before allowing the network to be put into normal operation,

- reviewing security logs, frequently with the aid of software audit tools to automate the log scanning procedure, and looking for signs of an intrusion or compromise, and
- performing an outside and independent audit on the normal operation of security features by systems administrators or others.

Usually, auditors are specially trained in IT cybersecurity techniques. One organization that trains IT cybersecurity auditors is the Information Systems Audit and Control Association (ISACA). Auditors with the certification ISACA sponsors, who are known as Certified Information Systems Auditors (CISA), are skilled in a variety of auditing methodologies for various IT systems and applications.

In a similar vein, an industrial network also needs a periodic audit to ensure that security countermeasures are set up, configured, and operating properly.

The goal of the industrial network security auditor is to find out if the countermeasures designed into the system are still operating effectively, the way they were designed and intended to operate, or if maintenance has fallen off and the countermeasures have not been updated, yielding an ineffective cyber defense.

## 8.6   Adding in Physical Security

As Chapter 2 emphasizes, physical security plays a major role in the security defense of any segment of the industrial plant, including the industrial network. Physical security countermeasures to prevent or deter unauthorized entry and/or access include measures such as locks on doors and windows, fences, and security guards. Countermeasures to detect unauthorized intrusions include burglar and intrusion alarms, closed-circuit TV (CCTV) cameras, and video recorders for those cameras. More recently there are video analytics software packages for

CCTV systems, which can alert operators to suspicious or unauthorized movements of people in restricted areas, etc. Physical security has been around for hundreds of years, and quite a number of sophisticated physical security devices are on the market.

There are many good sources of information on physical security in a plant environment. The American Chemistry Council (ACC) has a fair amount of material on physical security in its publication "Site Security Guidelines for the U.S. Chemical Industry."[3]

ASIS International, an international organization of security management professionals, has a wealth of good articles and resources on physical security on its web site[4], including articles from its monthly magazine, *Security Management*.

But perhaps the best advice on physical security for the industrial network security Program Team is also the easiest to follow: As urged in Chapter 2, include a representative of physical security or facilities management in risk assessment and other activities of the industrial network security Team. Without physical security representation, an important perspective will be missing.

## 8.7   Adding in Personnel Security

Like physical security, personnel security is another important component necessary to round out the industrial network security defense for an industrial plant. Some of the more common personnel security controls include the following:

- Background screening checks before hiring employees and contractors. These may include criminal record checks, credit checks, driving records, education records, etc.
- A clear statement of company security policies and the security behavior expected of employees and contractors.

- Company terms and conditions of employment, including measures such as employee rights and responsibilities and detailing offenses to security policies, disciplinary actions, etc.
- Incident investigation. Many big breaches of security are preceded by small breaches. All security related incidents should be investigated and the individuals involved monitored for indications of further security violations.
- Rechecking employees' and contractors' backgrounds periodically, especially after a security violation. This should be done in line with company personnel policies.

As with physical security, personnel security has been around a long time. There are many resources out there, and many practitioners. The previously mentioned ACC "Guide to Security at Fixed Chemical Sites" has a number of personnel security guidelines and recommendations. But, as mentioned previously in Section 8.6 regarding the field of physical security, the best advice the writer can give with personnel security is simply to have representatives of personnel security, whether the HR department or management or another group, sitting at the table when the risk assessment team or the industrial network security Program Team meets, and to make sure that their point of view is included.

# References

1. Computer Security Resource Center (CSRC) *Security Checklist for Commercial IT Products*. National Institute of Standards and Technology. Last updated 10/19/2004. Retrieved 11/11/2004 from http://csrc.nist.gov/checklists/.

2. Kirk, M. "Eligible Receiver" from PBS *Frontline* documentary "CYBER WAR!" Originally Broadcast 4/23/2003. Retrieved 11/11/2004 from http://www.pbs.org/wgbh/ pages/frontline/shows/cyberwar/.

3. American Chemistry Council, Chlorine Institute, and Synthetic Organic Chemical Manufacturers Association *Site Security Guidelines for the U.S. Chemical Industry.* 10/2001.

4. ASIS International Website. Retrieved 11/11/2004 from www.asisonline.org.

# 9.0
# New Topics in Industrial Network Security

## 9.1   Red Teaming: Test Yourself Before Adversaries Test You[1]

Red teaming traces its roots to warfare where commanders need to test and refine their own defenses and battle plans to ferret out weaknesses, study adversary tactics, and improve their strategies. Since this book covers industrial networks, our focus will be on cyber red teaming used to evaluate security questions related to these systems. Cyber red teaming has strong ties to both network vulnerability assessment and penetration testing.

Cyber red teaming, as you might expect, is a rather young field, but it is maturing as red teams have begun to collaborate, exchanging ideas, sharing tools, and developing new techniques. Over time, different groups have come to use cyber red teaming in one form or another, applying it to answer different questions (e.g., Are my personnel prepared to defend my network from a cyber attack? and Which of several security appliances will best protect my network?), and in different domains (e.g., cyber and physical).

But what exactly is red teaming? A key factor is that red teaming is *mission-driven*.

---

*1Note: Mr. John Clem of Sandia National Laboratories was a major contributor to the material in Sections 9.1 – 9.4.*

Many different groups perform red teaming and use differing terminology, techniques, and processes: commercial security firms, various military units and government agencies, and national laboratories. If one wants to understand a group that performs red team assessments then first one must understand what that group *means* by red teaming. For instance, Sandia National Laboratories' Information Design Assurance Red Team (IDART™) group defines red teaming as "authorized, adversary-based assessment for defensive purposes." The IDART group advocates that red team assessments be performed throughout any cyber system life cycle but especially in the design and development phase where cooperative red team assessments cost less, and critical vulnerabilities can be uncovered and mitigated more easily.

## 9.2   Different Types to Answer Different Questions

The IDART group has been red teaming for the U.S. government and commercial customers since 1996 and is widely known in the red team community. IDART identifies eight unique types of red teaming that can be performed individually or can be combined with other types. They are quick to point out that careful, detailed planning of a red team assessment requires significant communication between assessment customers and their red team. Experienced red teams should provide their customers with technical options for an efficient and effective assessment process that addresses their customers' security concerns.

The eight types of red teaming identified by IDART in their Red Teaming for Program Managers course are:

1. Design assurance (to improve new or existing system designs)

2. Hypothesis testing (to measure performance against a well-formed hypothesis)

3. Red team gaming (to evaluate adversary attack decision making in a given scenario)

4. Behavioral analysis (to analyze adversaries in order to identify indications and warnings)

5. Benchmarking (to produce a performance baseline that helps measure progress)

6. Operational (to test personnel readiness and defensive tactics, techniques, and procedures)

7. Analytical (to formally measure and compare available adversary courses of action)

8. Penetration testing (to determine whether and by what means an adversary can compromise system security).

# 9.3   Red Teaming Industrial Networks – Caution, It's Not the Same!

Most red teams don't assess industrial networks because they lack the specialized knowledge and training required to assess the sensitive components found in industrial networks. Industrial networks provide critical real-time or near real-time control over physical processes, and cyber red teaming sometimes results in intentional or accidental denials-of-service. Active network assessments (including penetration testing) should almost never be conducted in a production control system or control system network.

Where a control network interfaces with a business network, cyber assessment teams should be expert in understanding (and verifying) the network boundaries and how traffic is passed between the networks. Vulnerability scans and network foot-printing activities routinely executed by both network administrators and independent assessment teams in traditional IT networks can have extremely adverse impacts on industrial networks.

Instead of conventional active assessments, industrial network stakeholders must enable assessments (including red teaming) by using passive techniques and isolated test systems and networks. Still, integrating red team assessments into industrial network environments demonstrates an aggressive, proactive, security-conscious culture. The keys to success are what form of red teaming is implemented, who is on the team, and that a responsible, safe strategy is adopted to protect against accidental damage and/or disruption to the network.

## 9.4  System Security Demands Both Physical Security and Cybersecurity

Physical security systems are evolving to be increasingly dependent on cyber systems and information technology. For instance, physical access control systems at sensitive military, government, and commercial installations use computers, sensors, communications networks, databases, and other electronic information technology. Such security system networks are nearly indistinguishable from any other kind of IT network.

Indeed, new industrial network standards, such as those contained in NERC CIP, mandate physical security systems having greater capabilities. These systems contain functionality (like streaming video) that require bandwidth that is not found in a 24-Kb process control line, but which is found in a 100- to 1000-Mb business network.

One easy solution for network owners is to run the physical security communications through the business network, and perhaps establish a WiFi connection for remote sensors. The problem is that if someone is successful in compromising the business network, they are now within striking distance of the physical security system. Another approach might be to run some or all of the physical security system communications through the control systems network. In some instances this can work well, but in others it can represent a big risk to the control systems network.

The bottom line is, given the emerging trend in physical protection systems–incorporating COTS networking technologies and communications protocols–a capable adversary (outsider or insider) is but a stone's throw away from changing a physical security database and letting somebody inside a sensitive facility whom you *don't* want inside.

Because attacks against any kind of system or network can use physical means, cyber means, or both, a comprehensive approach to security requires assessments of both physical security and cybersecurity. Even more, system defenders must understand the concept of blended attacks, whereby an attacker uses physical means to enable cyberattacks, and cyber means to enable physical attacks. System owners and defenders should consider that cyber red teaming their industrial and administrative networks without also red teaming their physical security is inadequate.

Finally, performing red team assessments is not a task for amateurs. Even professional security organizations that lack specific experience in red teaming should consult with experienced red teams to consider a variety of assessment questions, options, recommended practices, legalities, and lessons learned before attempting to implement a red team assessment.

# 9.5   The Transportation Connection: Passenger Rail and Cybersecurity

By 2005 many industry sectors, such as oil and gas, chemicals, and electric power were already aware of, and working on, aspects of industrial network security. Much of the critical infrastructure in these sectors is privately owned; what about publicly owned infrastructure, such as in the transportation sector, particularly passenger rail?

The passenger rail industry in the United States has an interesting variety of systems. It contains some of the oldest and largest subway systems in the world, including New York City Transit. To that one may add showpiece subway systems like Washington, D.C.'s WMATA, new, sleek light rail systems such as Houston Metro, and advanced people-mover and commuter rail.

Passenger rail, as with other critical sectors mentioned earlier in this book, has not been without its cyber incidents. For instance:

- In 2003 a computer virus shut down the CSX system. Amtrak trains, which normally use the freight company's rails, were likewise shut down for hours.[1]
- In 2007 a 14-year-old Polish teenager in the city of Lodz hacked into the city's tram system, causing two streetcars to collide head-on and sending passengers to the hospital.[2]
- In 2006 in Toronto, a hacker changed the electronic passenger advertising on train signboards to display a disparaging comment about Canada's prime minister.[3]

In the summer of 2005, the writer approached APTA, the American Public Transportation Association, with a proposal. APTA is the trade association for North America's passenger rail and bus public transit agencies and associated industry. Public transit, covering everything from big city subways and commuter rail to newer light rail lines, was undergoing a change in control systems from old electromechanical relay and serial communications systems to modern industrial networks using PLCs, fiber optics, wide area networks (WANS), and Internet protocol (IP)-based communication. Would APTA be interested in jumping on the same bandwagon as the industries mentioned above, and support a control security initiative?

The writer recalls the meeting with APTA's staff at their Washington, DC headquarters: "I had the usual articles about control system security, concerning computer viruses and worms, and I was making moderate progress, when I decided it was time to pull out my heavy ammunition: a copy of *2600, the Hackers Quarterly*, Spring 2005 edition, freely available in many big bookstores.

This publication had a article on hacking the MetroCard® fare collection system, which is used by a number of big city subway systems. The author of the *2600* article had reverse engineered the information encoded on the magnetic stripes on these cards, and researched the original patents on the system to gain knowledge of the technical details. It was a full description of the system, how the cards are encoded (and how to decode them), how theoretically the cards could be overwritten (with a disclaimer to the effect that the author surely wouldn't want any of their readers to do anything illegal such as trying to change the amount stored on the cards and try to use them!). In all, the article was very professionally done, and would have made any technical editor proud."

That article did it! I had made a sale on the value of industrial network security to APTA. With some more awareness and organizational efforts, the APTA "Control and Communications Security Working Group" was created and funded. At the time of this writing, Part 1 of the Recommended Practice "Securing Control and Communications Systems in Transit Environments" is in the balloting/approval stage. Part 1 contains getting organized and background information for transit agencies, up through risk assessment. Part 2 will follow, which will contain developing a security plan and designing, installing, and maintaining security controls.

# References

1. Hancock, D. "Virus Disrupts Train Signals." *CBS News.com* article, 8/21/2003. Retrieved 8/2/2009 from http://www.cbsnews.com/stories/2003/08/21/tech/main569418.shtml.

2.	Leyden, J. "Polish Teen Derails Tram after Hacking Train Network." *The Register*, 1/11/2008. Retrieved 8/2/2009 from http://www.theregister.co.uk/2008/01/11/tram_hack/print.html.

3.	Leyden, J. "Hackers Libel Canadian Prime Minister on Train Signs." *The Register*, 5/3/2006. Retrieved 8/2/2009 from http://www.theregister.co.uk/2006/05/03/canadian_train_sign_hack/.

# 10.0
# Defending Industrial Networks—Case Histories

## 10.1 A Large Chemical Company

In this section, we will take a look at a case history of a large multinational corporation in adding industrial network security to its control networks.

The figures we will use to illustrate this story have been taken from slides given by this company at a past conference.

Figure 10-1 shows the typical situation in the company as far as industrial networks were concerned before the industrial network security push.

Here, we see that the business LANs and the process control network (the Process Control LAN in the diagram) were blended together, making up a corporate Intranet.

The revised network architecture, after an intensive campaign to isolate the process control network, is shown in Figure 10-2. The "E-Pass" notation on the diagram will be explained later in this section.

Here we see a complete reengineering to separate the business LAN, or Intranet, from the Process Control Network (PCN). If we refer back to Chapter 6, the design and planning philosophy of defense in layers was applied to separate the business LAN and the Process Control Network using a firewall.

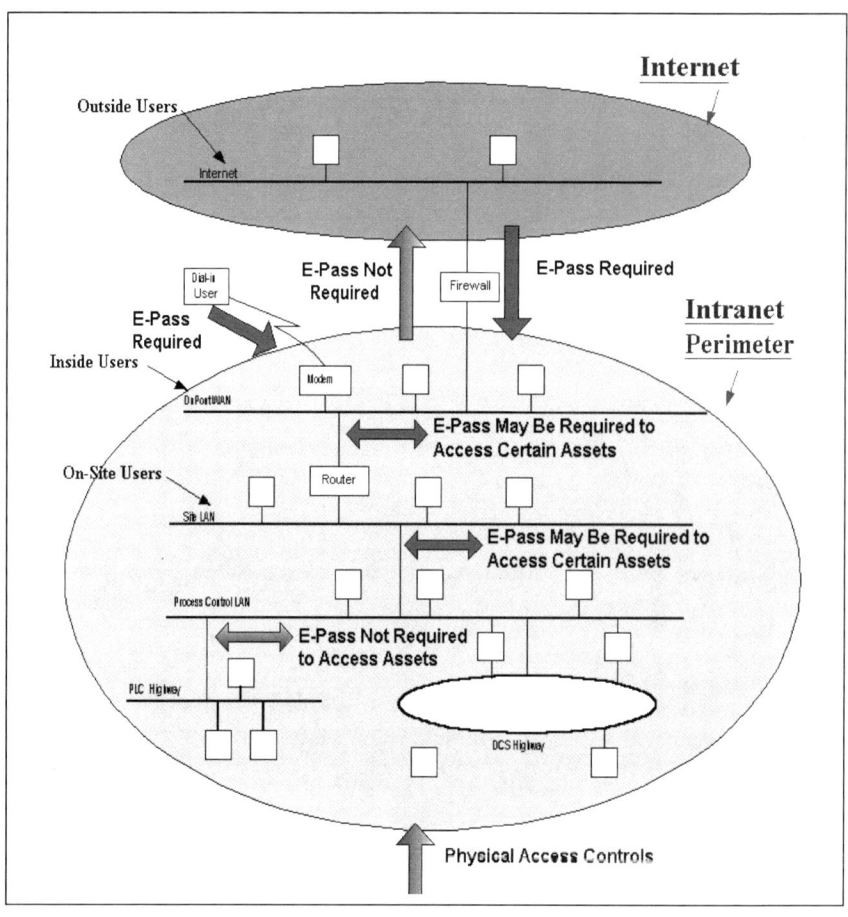

**Figure 10-1. Pre-Existing Security Controls**
**Note – E-Pass = Two Factor Authentication (RSA)**

Figure 10-3 shows how several firewall options were tried by the company, and the low-cost "SOHO" type appliance (single office/home office) was rejected. A moderate-size enterprise level firewall was selected.

It is important to mention that the company did not attempt to do this internal firewall addition/network separation exclusively in-house. Rather, the company chose to partner with a Managed Firewall Provider, an external vendor that supplied the firewalls and provided off-site monitoring and firewall expertise for the company's plant networks

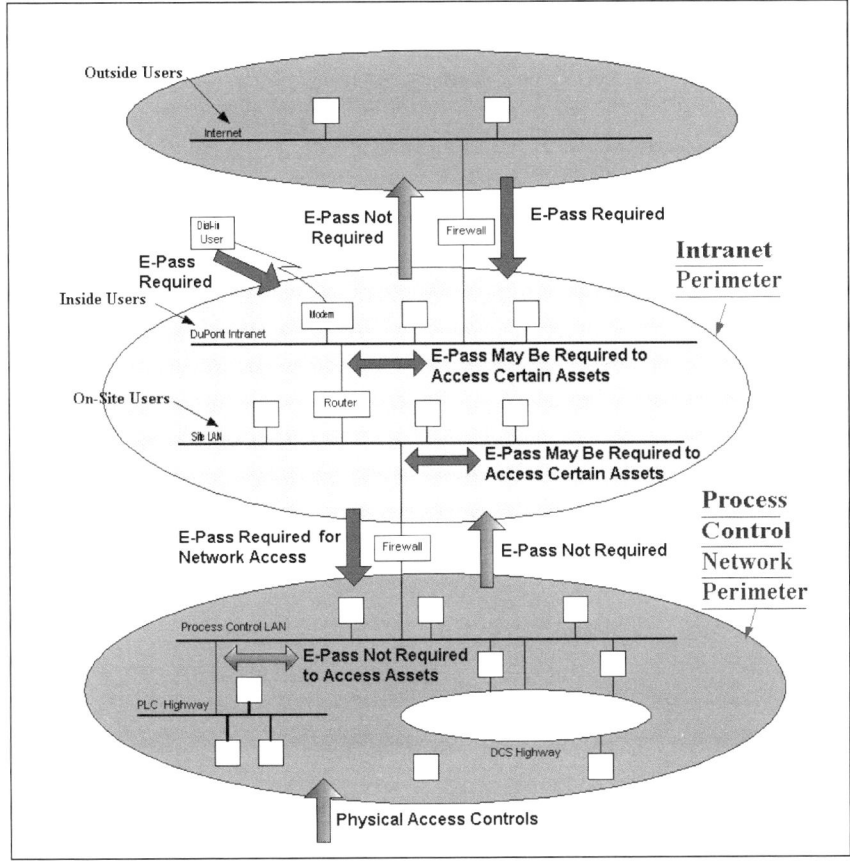

**Figure 10-2. New Perimeter-Based Security Controls**

around the world. The Managed Firewall Provider concept is used in the business world by many medium and large companies that do not want to do the entire job in-house.

Figure 10-4 shows how communication typically flows across the internal firewall from the "clean" process side to the business side for such things as backups, OPC data updates, antivirus signature file updates, and so on.

Figure 10-5 gives a performance summary, based on the number of installed firewalls (more than 60). As the figure mentions, the necessary

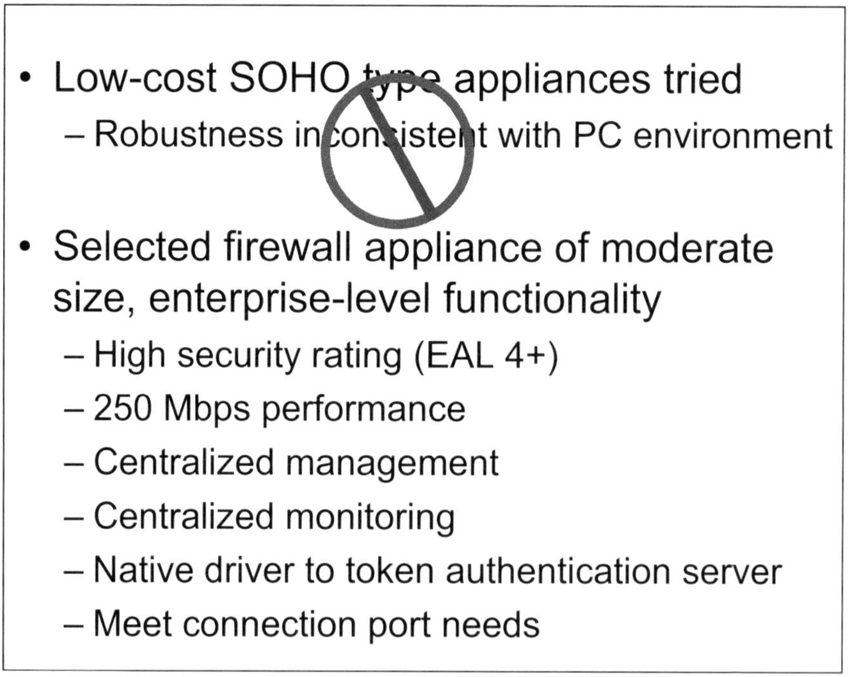

- Low-cost SOHO type appliances tried
  - Robustness inconsistent with PC environment

- Selected firewall appliance of moderate size, enterprise-level functionality
  - High security rating (EAL 4+)
  - 250 Mbps performance
  - Centralized management
  - Centralized monitoring
  - Native driver to token authentication server
  - Meet connection port needs

**Figure 10-3. Firewall Characteristics**

process communications were handled with no throughput issues, and the conclusion is that "standard IT firewall technology *can* be used for process control applications".

Let's now turn our attention to the caption "E-pass" that is mentioned in Figures 10-1 and 10-2. E-Pass is a two-factor remote access authentication method used corporate-wide at this company. The technology is supplied by a commercial cybersecurity provider, RSA. As you will notice in Figures 10-1 and 10-2, the diagrams mention "E-Pass Required," or "E-Pass Not Required," or "E-Pass May be Required to Access Certain Assets."

The RSA token-based, two-factor authentication scheme uses a centralized server that is queried to securely authenticate that remote users are who they say they are. Access rights to hosts on the network are provided by the applications and/or internal process control firewall.

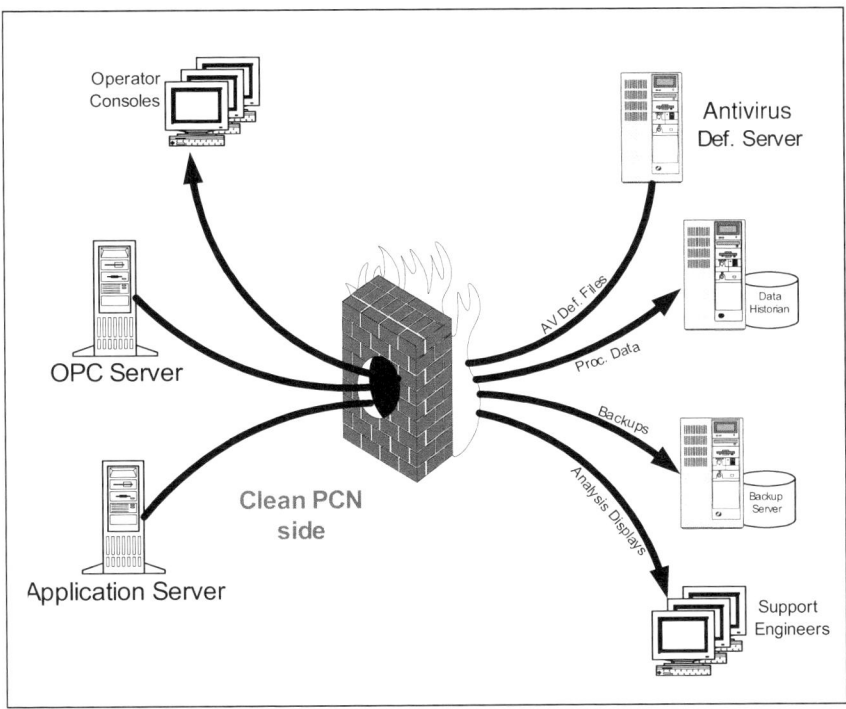

Figure 10-4. Typical Communications

- **Based on 60+ installed firewalls:**

  - Adequately handle process communications

  - No through-put issue with standard IT firewall appliance device (appropriately sized and integrated into the architecture)

- *Standard IT technology firewall <u>can</u> be used in process control applications*

Figure 10-5. Performance

To summarize, this case history shows that a large corporation with plants across the globe was able to very successfully apply some fundamental strategies of industrial network security and separate their Process Control Networks off with firewalls.

## 10.2 Another Company's Story–Procter & Gamble

In this section, we will look at a case history from a second large corporation, Procter & Gamble. This time we will focus on how a large company views industrial network security risks and performs a qualitative risk analysis, as was described in Chapter 2. The figures to illustrate this story were provided by Dave Mills, a Technology Leader in Procter & Gamble's Corporate Engineering organization.

Figure 10-6 shows a general model for developing a risk management process for emerging areas of risk. At Procter & Gamble, this model was helpful, but reality proved more complicated. In order to obtain the human resources to perform the qualitative risk assessment, an initial screening assessment was needed to persuade management that a more in-depth study was justified. The Risk Reduction Program appears fairly linear in Figure 10-6, but, in reality, the security goals and standards were developed in parallel with the security controls. If you are developing a risk management program while you are experiencing the risks, you often don't have the time to perform each step in series.

Dealing with risk is not a new phenomenon at Procter & Gamble or other large corporations. Risk in more traditional and familiar areas has been analyzed, evaluated, and managed for years. What is new are the unique security risks associated with modern industrial networks and how to bring that risk "into the fold" alongside other risk management programs.

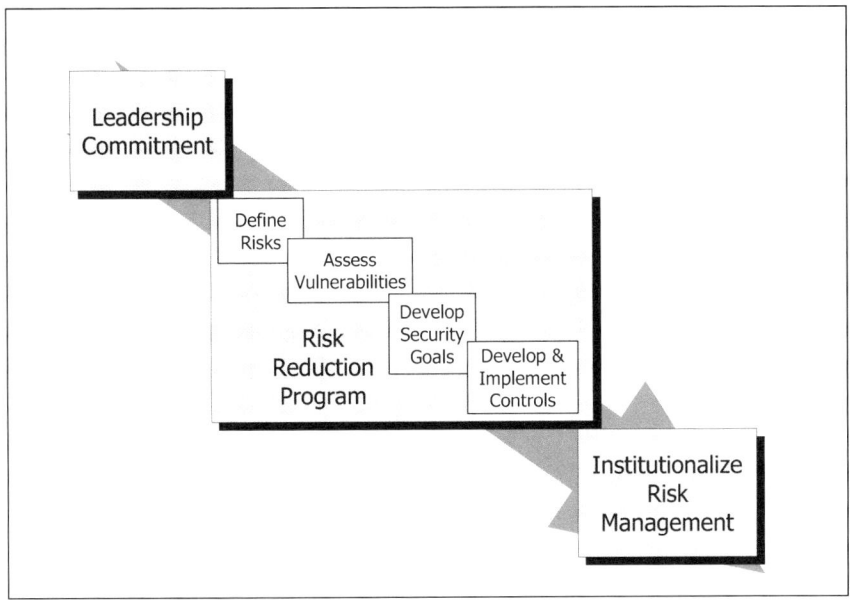

**Figure 10-6. Background-Risk Management (*Courtesy of Procter & Gamble*)**

Figure 10-7 shows the existing risk disciplines that industrial network security cuts across at P&G: Business Continuity Planning (BCP), IT Security (IT) and Health, Safety and Environment (HS&E).

Figure 10-8 shows how Procter & Gamble wound up with a specific risk assessment methodology: Facilitated Risk Assessment Process (FRAP). The primary customer was the Information Security organization, and this was the methodology they had the most experience with.

One of the main points Dave Mills stressed is that the whole risk assessment discussion is by nature different for different companies, as different companies have unique products, manufacturing locations, manufacturing hazards, and probably differing threat profiles. On the "soft" side, corporate culture and personnel management issues must be taken into account when performing an industrial network security risk assessment that matches your company.

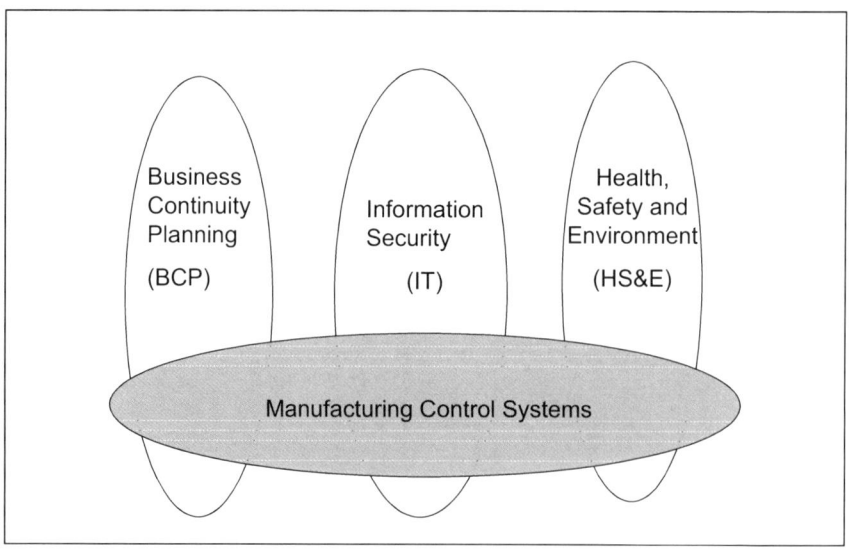

Figure 10-7. Risk Areas by Discipline (*Courtesy of Procter & Gamble*)

- Risk = Probability x Severity
- Dozens of Specific Methodologies
  - Cybersecurity VAM Process, CIDX, (www.cidx.org)
  - Approach to Industrial Cyber Security Vulnerability Analyses, Primatech, Paul Baybutt (www.primatech.com/technical/index.html)
  - Facilitated Risk Assessment Process, Tom Peltier & Associates, Tom Peltier (www.peltierassociates.com)
- Customers: Info Security, BCP, HS&E
- Customer Preference: FRAP

Figure 10-8. Risk Analysis Methodologies
(*Courtesy of Procter & Gamble*

*Many thanks to Dave Mills and Procter & Gamble Engineering for allowing their story to be published.*

# Appendix A – Acronyms

ACC      American Chemistry Council

AIC      Availability, Integrity, and Confidentiality

AIChE      American Institute of Chemical Engineers

AWWA      American Water Works Association

BCIT      British Columbia Institute of Technology

BPCS      Basic Process Control System

CCPS      Center for Chemical Process Safety

CIDX      Chemical Industry Data Exchange

CIO      Chief Information Officer

CISA      Certified Information Systems Auditor

CISSP      Certified Information System Security Professional

COTS      Commercial Off The Shelf

DCS      Distributed Control Systems

DHS        Department of Homeland Security

DoE        Department of Energy

FERC       Federal Energy Regulation Commission

GAO        General Accounting Office

GUI        Graphical User Interface

HMI        Human Machine Interface

IDE        Intelligent Electronic Device

M&CS       Manufacturing and Control Systems

NERC       National Electrical Reliability Council

NIST       National Institute of Standards and Technology

NISCC      National Infrastructure Security Co-ordination
Center

NRC        Nuclear Regulatory Commission

OCIPEP     Office of Critical Infrastructure Protection and
           Emergency Preparedness

OPC        Object Linking and Embedding for Process
           Control

PCSRF        Process Control Security Requirements Forum

PLC          Programmable Logic Controllers

SCADA        Supervisory Control and Data Acquisition

SIS          Safety Instrumented Systems

SPDS         Safety Parameter Display System

TCP/IP       Transmission Control Protocol/Internet Protocol

# Index

DoE National Laboratories 7
drawbacks 31

electronic security 3
email 19, 95
encryption 78, 85
    ciphertext 79
    encryption algorithm 79
    encryption key 79
    plaintext 79
E-pass 115, 118

facilitated risk assessment process
    (FRAP) 121
Federal Energy Regulation Commis-
    sion (FERC) 7, 9
firewall 63, 70–71, 116–118
for-profit entities 8

General Accounting Office (GAO) 7
government organizations 7

hacking websites 45
hardware 29
hardware/software system integrity
    35
host intrusion detection system
    (HIDS) 73

IACS 1–2
identification 36, 86–87
industrial automation and control
    systems (IACS) 1–2
industrial networks 1, 33
    case history 115
    defense 54
    security 1–2, 12
Information Design Assurance Red
    Team (IDART™) 108
infrastructure protection, critical 4
insiders 11
intent 40
internet 29
intranet 115
intrusion detectors 73

host-based 73
    network-based 73
ISA99 1, 3–4, 8
ISA-99 TR1 69, 72
IT cybersecurity 71, 94
    manager 18
IT security
    policies 19
    procedures 19
    processes 19
    technology 19

key management 85

LAN 63, 115
layered defenses 58
layers of protection 58
leadership commitment 93
least privilege 65
likelihood 40, 49
logic bomb 45

malicious acts 3
management actions 93
mandatory access control 90
manufacturing and control systems 8
message integrity checking 84

National Cyber Security Division
    (NCSD) 7
National Institute of Standards and
    Technology (NIST) 101
National Security Agency (NSA) 101
network
    architecture 115
    hardening 101
    separation 66
    sniffer 73
    traffic anomalies 73
network intrusion detection system
    (NIDS) 73
networking 41
    products 29
    protocols 29
nonprofit organizations 8

threat 39, 49
  cyber 40
  matrix 41
total cost of ownership 31
trade secrets 36
trained personnel 19
trojan horse 44, 63

unlicensed software 19, 95
utilities 1

virtual private networks (VPNs) 85
virus control 75
viruses 44
VPN configuration 86
vulnerability 40, 49
vulnerability assessment 107

web 7, 19, 95
web services 7
wireless 7
wireless LANs 7, 29
worms 44

zone and conduit 64